This third book in the *Rocks & Fossils* series covers the geology of a popular National Park close to several large conurbations. Like its companion titles, it will provide the student and enquiring general reader with an authoritative, concise and portable field companion to their enjoyment of part of Britain's countryside.

Scenically, the Peak District is one of the most attractive and least spoilt areas of Britain. Geologically, it is of great interest, with its limestone formations yielding a wide range of fossils and minerals. These qualities, together with its accessibility, make it an ideal area for studies of the natural environment.

As in other books in the series, the first part presents a general introduction to the geology of the area. The second part contains 15 itineraries in localities where geological features are particularly well displayed. The book is fully illustrated with maps, diagrams and photographs, and there is a glossary of specialised terms.

ROCKS AND FOSSILS
Editor: J. A. G. Thomas

*About the author*
Dr I. M. Simpson is Senior Lecturer in Stratigraphy at the University of Manchester. He lives near the Peak District and has spent much of the past 30 years doing research and leading field excursions in the area.

Every effort has been made to ensure that the information presented in this book is accurate and up to date. However, readers are invited to draw our attention to any inaccuracies − particularly in the information concerning access to land covered in the excursions − by writing to the author, c/o Unwin Paperbacks, PO Box 18, Park Lane, Hemel Hempstead, Herts HP2 4TE.

# The Peak District

## I. M. SIMPSON

London
**UNWIN PAPERBACKS**
Boston          Sydney

First published in Unwin Paperbacks 1982

Unwin® Paperbacks,
40 Museum Street, London, WC1A 1LU

Unwin Paperbacks,
Park Lane, Hemel Hempstead, Herts. HP2 4TE

George Allen & Unwin Australia Pty Ltd.,
8 Napier Street, North Sydney, NSW 2060, Australia

DGV
emv
S

**British Library Cataloguing in Publication Data**

Simpson, I. M.
 The Peak District. − (Rocks and fossils; 3)
 1. Geology − England − Peak District
 I. Title    II. Series
 554.25′11       QE262.P/
 ISBN 0-04-554006-3    81   010345

Set in 9 on 11 point Times by Typesetters (Birmingham) Ltd.,
and printed in Great Britain
by Hazell Watson & Viney Ltd, Aylesbury, Bucks

# Foreword

Over the past few years there has been increasing public interest in the geological sciences. This derives from an increased awareness of their twofold role: first, they sketch out the history of our planet, and secondly they help to provide the mineral resources on which modern society depends. Although this spread in interest is to be welcomed, it can lead to certain undesirable consequences as more and more visitors come to examine a finite number of instructive rock exposures.

The author and publishers of this book are fully aware of the need to make the best possible use of the outcrops it describes and have accordingly consulted with the Nature Conservancy Council, the official body responsible for conservation in Britain. All their efforts however will be nullified if today's readers choose to ignore their responsibility to students of the future.

It must be emphasised that the vast majority of geological outcrops are in private hands and that access to them is through the goodwill of the owners and occupiers. If lost, this goodwill will be difficult, if not impossible, to regain — it can take only one careless act to cause offence. Further, many geological localities can lose their interest through the cumulative effects caused by the unnecessary use of hammers. To observe and record, collecting any necessary specimens only from fallen rock, will in general give a better understanding of geology than will a physical assault on selected portions of the rock face. The indiscriminate use of hammers is thus as profitless as it is damaging.

The author and publishers of this guide have done their best to ensure the maximum benefit from your visit to the Peak District; it remains for you to ensure that the same benefit remains available to your successors.

DR G. P. BLACK
*Nature Conservancy Council*

# Acknowledgements

The author is greatly indebted to his friends, colleagues at the University of Manchester, the Institute of Geological Sciences at Leeds and the Peak Park Joint Planning Board at Bakewell for all the help, advice and encouragement they gave him while he was writing this book.

The maps are based in part upon the Ordnance Survey map with the permission of the Controller of Her Majesty's Stationery Office, Crown copyright reserved.

# Contents

# GEOLOGICAL TIME SCALE

| Era | Period | | Age to base, Ma |
|---|---|---|---|
| CAINOZOIC | Quaternary | Holocene (Recent) | 0·01 |
| | | Pleistocene | 1·8 |
| | Tertiary | Pliocene | 6 |
| | | Miocene | 23 |
| | | Oligocene | 38 |
| | | Eocene | 55 |
| | | Palaeocene | 65 |
| MESOZOIC | Cretaceous | | 140 |
| | Jurassic | | 195 |
| | Triassic | | 230 |
| PALAEOZOIC (UPPER) | Permian | | 280 |
| | Carboniferous | | 345 |
| | Devonian | | 395 |
| PALAEOZOIC (LOWER) | Silurian | | 435 |
| | Ordovician | | 500 |
| | Cambrian | | 570 |

The three eras above may be grouped together as the Phanerozic.

Precambrian time is divided as follows:

| | |
|---|---|
| Proterozoic | 2600 |
| Archaean | |
| Age of the Earth | 4500 |

The ages given are approximations

# Part 1: Geological background

# 1 Introduction

Fifteen one-day and half-day geological field excursions are described in this book. The selection of these from an area as large and varied as the Peak District was not a simple task, for there are well over a hundred possible excursion routes to choose from. Three major factors governed the final choice. The first was the necessity to have the locations of the excursions spread as widely over the area as possible; secondly, it was desirable to include as great a variety of geological phenomena as possible. The third factor was ease of access to the sites of interest: wherever possible the excursion routes lie along public footpaths, bridle paths, trails and minor roads. In this way the worst of the road traffic is avoided, and the tiresome and often futile task of writing in advance for permission to cross private land or to enter quarries and mines is eliminated.

Several of the excursions, such as those in the Castleton and Dovedale areas, are in parts of the Peak District which are very popular with many visitors other than geologists and are, therefore, liable to become quite crowded at times, especially summer weekends. In contrast, some of the excursions are on much less-frequented routes where, quite possibly, you might not meet another person all day.

Inevitably the author's choice of excursions has been influenced by his own personal preferences. Excursion 7 is included because it is his firm belief that a visit to Lathkilldale on a sunny day, midweek in May, comes as near to perfection as any geological outing anywhere in the world possibly could.

When writing this book it has been assumed that readers will already have some basic knowledge of geology and be reasonably familiar with the more common rocks, minerals and fossils. Beginners need not take fright, however. The author's intention has been to describe the geology of the region in as simple and straightforward a style as possible and to restrict to a minimum the use of long-winded technical and unfamiliar words, which so bedevil the subject of geology for so many students. A short glossary of those terms whose use could not easily be avoided is provided at the end of the book.

Essential to the full enjoyment of the excursions are good topographic maps of the area. The two Ordnance Survey 1:25 000 Outdoor Leisure Maps of the area — The Dark Peak and The White Peak — are strongly recommended.

The Peak District National Park Joint Planning Board provides a very wide range of useful information about the area, such as details of interesting walks, access to open country, accommodation, bus and train timetables, special events, and so on. Anyone planning a stay in the area would find it well worthwhile writing to the Joint Board at their office at Aldern House, Baslow Road, Bakewell, Derbyshire.

For no good reason whatever some geologists take great delight on field excursions in cracking rocks by hitting at them with hefty blows from specially designed hammers. In the Peak District it is quite unnecessary to do this. In fact, the area lies almost wholly within the confines of the National Park where the hammering of rocks is strictly, and rightly, prohibited in the interests of conservation of the countryside.

So, if you are a compulsive hammerer of rocks, and cannot rid yourself of the habit, please leave your hammer at home when you come on any of the excursions. By so doing you will not only enjoy the bonus of having an appreciably lighter rucksack to carry but also, by using your eyes instead of your hammer, you will find that there is a lot more to see than you ever imagined.

# 2 Geological history of the Peak District

Almost all the rocks exposed at the surface in the Peak District belong to the Carboniferous System. They were formed between 280 and 345 million years ago at a time when Britain lay much closer to the Equator than it does now. Three of the major subdivisions of the system − the Carboniferous Limestone (or, to use its alternative name, the Dinantian) Series, the Millstone Grit (or the Namurian) Series and the Coal Measures (or the Westphalian) Series − are represented, the first and second of these more fully than the third. At the close of the Carboniferous Period the rocks were subjected to moderately strong folding and faulting followed by uplift and much erosion. The present surface distribution of the rocks is shown in Figure 1.

All the rocks of the Carboniferous Limestone Series seen at the surface belong to the upper (Viséan) subdivision of the series. They are predominantly limestones but some shales, mudstones and rocks of volcanic origin also occur. It is thought that, for most of the time during which the rocks of the Carboniferous Limestone Series were being deposited, a shallow, warm sea covered most of the area.

Three principal types, or **facies**, of limestone can be recognised. Each type is indicative of a particular type of depositional environment. The type that occurs most commonly is the **shelf limestone**. This was deposited as lime-rich sand and mud on the bottom of a very shallow part of the sea which covered the area. The sea may have been no more than a metre or two in depth.

Shelf limestone is a fine-grained, hard, brittle rock. Generally it is pale grey in colour. It usually occurs in clearly stratified beds which vary from thin bands only a few centimetres thick to massive layers several metres in thickness.

Shelf limestone is composed mainly of fragmented pieces of the shells and **calcareous** skeletons of various marine organisms held together by a cement of finely crystalline **calcite.** Some of the beds display **oolitic** texture, but this is rare. Large, unbroken fossils are conspicuously absent from many of the beds, but when they do occur they are usually abundant. The most commonly occurring fossils are **brachiopods, corals** and **crinoids.** Beds of shelly limestone containing enormous numbers of the large **productid** brachiopod, *Gigantoproductus* (Fig. 2), occur at several different horizons in the succession. Beds crowded with colonies of the compound coral, *Lithostrotion*, also occur occasionally.

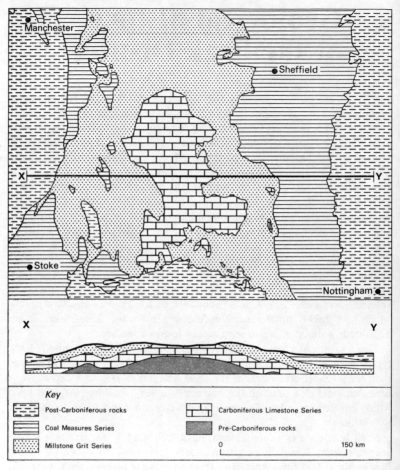

**Figure 1** Geological sketch-map of the Peak District and adjoining areas, with a diagrammatic cross-section to illustrate the structure.

The second most common type of limestone is **basin limestone**. This type is darker in colour than shelf limestone and it tends also to be more thinly bedded. The individual beds are often separated from one another by partings of **shale** or **mudstone**. Fossils are scarce in some beds, abundant in others. Brachiopods, corals and crinoids again are the common fossils. Basin limestones contain more clay sediment than do shelf limestones and they are thought to have been deposited in deeper, muddier water.

The third type of limestone to occur in large quantity is **reef limestone**. Typically it consists of limestone in which stratification is either poorly

**Figure 2** *Gigantoproductus* in Ricklow Quarry, Lathkilldale (× 1).

developed or non-existent. Reef limestone is frequently very rich in fossil remains (Fig. 3). Calcareous **algae** – plants similar to modern seaweeds whose tissues, during life, become coated with a deposit of lime to form hard skeletal frameworks – are very common. Many different types of animal fossils occur. These include brachiopods, **bryozoans**, corals, crinoids, **gastropods, goniatites, ostracods** and trilobites. The brachiopods

**Figure 3** Fragments of crinoids and brachiopods in reef limestone from Cave Dale (× 1).

are especially common and consist mainly of productid, **rhynchonellid**, **spiriferid** and **terebratulid** species.

Reef limestone occurs either as isolated patches within the basin and shelf limestones or as a marginal deposit between these two types. The outcrops vary in size from small areas only a metre or two wide to very large structures several hundreds of metres across. It is thought that reef limestone was formed in much the same way as patch and fringing reefs are being formed at the present time in shallow tropical seas rich in marine life.

The most extensive exposures of reef limestone are in the marginal reefs in the vicinity of Castleton in the north and Dovedale in the west of the area. In these places the reef limestones reach a maximum thickness of about 500 m.

The shelf limestones reach their fullest development in the central part of the area where they are about 550 m thick. By using minor variations in the **lithology** of the limestones, the succession can be subdivided into a sequence of local formations (Table 1).

The basin limestones, best developed in the west of the area, where they are about 750 m thick, can likewise be subdivided into several lithological formations. The correlation of these formations with those in the shelf limestone and with the standard fossil zones and stages of the Lower Carboniferous succession of north-west Europe are given in Table 1.

At various intervals during the deposition of the Carboniferous Limestone Series, volcanoes were active in the Peak District. Molten lava

**Table 1**   The succession in the Carboniferous Limestone Series.

| Basin facies | Shelf facies | Fossil zones | Stages |
|---|---|---|---|
| Widmerpool Formation and Mixon Limestone–Shales | Longstone Mudstones | $P_2$ | Brigantian |
| | Eyam Limestones | | |
| | Monsal Dale Limestones | $D_2$ | |
| Ecton and Hopedale Limestones | Bee Low Limestones | $D_1$ | Asbian |
| | Woo Dale Limestones | $S_2$ | |
| Milldale Limestones | Iron Tors Limestones | $S_1$ $C_2$ | Arundian |

and ash were erupted from several volcanic vents. During the larger eruptions very extensive areas of shelf limestone were covered by lava flows which, on cooling, solidified to form beds of **olivine-basalt** up to 30 m thick.

The lava flows are easily distinguished from the limestones, in which they are interbedded, on account of their much darker colour. Specimens of **basalt** from different flows may, however, differ widely in appearance when examined in detail. Some are hard, smooth, crystalline rocks, dark greenish-grey in colour. However, the crystals are too small to be easily identified, even with a magnifying lens.

Other lavas, when erupted, contained large amounts of dissolved gas and steam which, when escaping from the lava, caused it to froth and foam. The bubbles which were trapped inside the lava as it solidified are preserved in the rock as small spherical cavities. Basalts with this 'frothy' texture are said to be vesicular. Very often the cavities, or vesicles, were later filled with secondary minerals such as calcite and quartz. These infillings are known as **amygdales**, and the texture of the rock is then said to be amygdaloidal.

Basalt is very susceptible to chemical weathering. The iron-rich silicates in the rock undergo oxidation and hydration; the feldspars are leached of their sodium and calcium. The end-product of this process is a mixture of soft clay minerals and rust-brown iron hydroxides. Partly decomposed basalt is a soft, crumbly, mottled greenish-grey and brown rock (known locally as **toadstone**); it is very different in appearance from fresh, unweathered basalt.

Minor eruptions of volcanic ash took place frequently in the Peak District. As a result of this, many of the beds of limestone are separated from one another by thin layers of pale greenish-grey clay. These layers of weathered and partially decomposed volcanic ash, rarely more than a few centimetres thick, were termed **'clay wayboards'** by the Derbyshire lead miners. Sometimes rich deposits of lead ore were found in the limestones immediately underneath the clay wayboards.

Not all the volcanic activity was confined to surface eruptions of lava and ash. At some depth below the surface molten magma forced its way in between some of the beds of limestone to form intrusive **sills**. The magma in the sills, cooling more slowly than the lava flows on the surface, solidified to form **dolerite**. This is similar to basalt as far as its mineral content is concerned but is visibly coarser in grain size.

Like basalt, dolerite frequently undergoes alteration as a result of chemical weathering. Outcrops of the rock are often stained rusty brown due to oxidation of the iron silicates in the rock, and in extreme cases these oxidation products form a thick skin around solid blocks of fresh dolerite.

The limestones, too, undergo secondary alteration in several ways. In some of the shelf limestones, for instance, some of the calcium in the rock has been replaced by magnesium, thereby converting the limestone to **dolomite** rock. This is generally more obviously crystalline than the limestone it replaces and often is pale yellowish-brown in colour.

Another, more widespread, change is the partial replacement of the calcium carbonate in both the shelf and basin limestones by silica in the form of the mineral, **chert**. This occurs as hard, black nodules embedded in the limestone and, as they are much more resistant to weathering and erosion than the limestones, they stand out prominently on exposed surfaces.

Because of the ease in which calcium carbonate dissolves in slightly acid groundwater, all the various types of limestone have undergone solution to some extent by water percolating through any cracks and fissures in the rocks. This process, continued over millions of years, eventually led to the development of complex underground cave systems on a large scale. Frequently the calcium carbonate, having been dissolved in one part of the area, is transported in solution to another part where it may be precipitated out of solution in the form of **stalactites**, **stalagmites** and **flowstone** on the sides of fissures and in caves. When water in which calcium carbonate has been dissolved emerges at the surface from underground as a spring, the lime is frequently precipitated in the form of calcareous tufa, a soft, crumbly deposit which encrusts the rocks, pebbles and even the vegetation in the vicinity of the spring.

The highest beds in the Carboniferous Limestone Series consist of dark grey, muddy limestones and mudstones in which **bivalves**, goniatites and ostracods are common fossils. This change to a muddier environment of deposition foreshadows the change to more turbulent conditions that were to follow during the deposition of the Millstone Grit Series.

The Millstone Grit Series follows the Carboniferous Limestone Series conformably in some parts of the Peak District, but in other parts there was a period of uplift and erosion between the two periods of deposition so that the upper series, locally, lies on the lower series with strong **unconformity**. In the vicinities of Buxton and Castleton, for example, the basal beds of the Millstone Grit Series lie on an eroded surface of the Bee Low Limestones with about 200 m of the uppermost strata of the Carboniferous Limestone Series missing.

Mudstones, shales and sandstones are the predominant rock types found in the Millstone Grit Series. The shales and mudstones are relatively soft, dark-grey rocks composed principally of very fine particles of clay minerals and quartz. Even with a powerful magnifying lens, the particles are too small to be identified. Shale, unlike mudstone, can be split readily into thin sheets along the bedding planes. This property is due to the presence of thin flakes of mica and similar minerals lying flat along the bedding planes.

When exposed to the weather, shales and mudstones disintegrate readily and eventually revert to their original state — soft, black mud.

Nodules of irregular shape, ranging in size from a few centimetres across up to one metre, often occur embedded in the shales and mudstones. As they are much harder than the surrounding rock, they stand out prominently on exposed surfaces. The nodules, known as **bullions**, consist of patches of the original muddy sediment which, soon after deposition, became strongly cemented with secondary deposits of iron and calcium carbonates, presumably precipitated from circulating groundwater. Well preserved fossils such as bivalves and goniatites, can often be found inside the bullions (Fig. 4).

**Figure 4**   A goniatite from the Mam Tor area (× 1).

Most of the sandstones in the Millstone Grit Series are coarse grained; when handled they feel distinctly rough and gritty. Surfaces which have been exposed to the weather for a long time are very dark, almost black, in colour, but freshly exposed surfaces are in pale shades of brown, often with a pink or yellow tinge. Quartz, in the form of angular grains, is the principal constituent. Flakes of mica are common in many examples. Small patches of pink or white clay can often be seen amongst the quartz grains. These were originally grains of feldspar but, after deposition they underwent chemical weathering and decomposed to their present state.

In the more coarse-grained varieties of sandstone the sand grains are clearly visible. Small, rounded pebbles of white quartz occur frequently in these coarse rocks. Well defined **cross-stratification** (cross-bedding) is a conspicuous feature of most of the sandstones (Fig. 5).

The succession in the Millstone Grit Series begins with a sequence of dark-grey shales and mudstones. In the north these are known as the Edale Shales and they reach a thickness of 200 m. Towards the south-east of the area these basal beds thin to less than 50 m. On the west side, thin beds of hard sandstone and siltstone are interspersed among the shales. Marine

**Figure 5**   Cross-stratification in the Ashover Grit, Youlgreave.

fossils, such as goniatites, occurring in the basal shales indicate that deposition of the sediment took place in a muddy sea.

These basal shales and mudstones are followed by thick beds of sandstone. Deposition of these was at first confined to the more northern parts of the area and begins with the Mam Tor Beds and Shale Grit which reach a maximum thickness of nearly 300 m in the Edale Valley. More mudstones and shales followed, and then the Kinderscout Grit, which consists of thick beds of medium- and coarse-grained, pebbly sandstones was deposited. It has a maximum thickness of 150 m in the north of the area but thins rapidly southwards.

The thick, gritty sandstones of the Shale Grit and Kinderscout Grit are thought to have been deposited on the frontal slopes of a large delta at the mouth of a river which drained an extensive area of land further north. The intervening mudstones and shales represent a prolonged incursion of muddy seas over the delta.

The Kinderscout Grit is followed by more marine shales and mudstones. These in turn are succeeded by the Ashover Grit in the south-east of the area and the Five Clouds Sandstone and Roaches Grit in the south-west. Because these sandstones thin and eventually die out when traced northwards, but thicken and become coarser to the south, a southerly source is inferred.

Marine shales overlie the Ashover Grit, but these are soon followed by a coal seam, the first of many in the succession from now on. These coal seams indicate that, by now, much of the delta lay above sea level for periods lengthy enough to allow extensive covers of peat-forming vegetation to develop.

The two uppermost sandstone formations in the Millstone Grit Series are the Chatsworth Grit and the Rough Rock. Both are widespread over the area but are thicker and coarser in the north-east, suggesting a derivation from that quarter.

A thin band of marine shale bearing the fossil goniatite, *Gastrioceras subcrenatum*, is the dividing line between the Millstone Grit Series and the Coal Measures Series, which follows without a break. The rocks in the Coal Measures Series are very similar to those in the upper part of the Millstone Grit Series. Mudstones, shales, sandstones and thin coal seams predominate, but the sandstones are less coarse and the coal seams more numerous than in the underlying series. Plant fossils commonly occur (Fig. 6).

Only the lowermost 150 m or so of the Coal Measures Series is seen in the area covered by this guide; it includes one thick sandstone, the Woodhead Hill Rock, and the Yard Coal, a seam thick enough to have been mined quite extensively at one time.

**Figure 6** *Stigmaria*, a fossilised tree root in Coal Measures sandstone at Goyt's Moss ( × ½).

The greater part of the Coal Measures Series that once covered the entire area was removed during long periods of erosion in post-Carboniferous times. This erosion was a consequence of the general rise in land level that took place in Britain, first at the end of the Carboniferous, and again, after submergence below sea level in the Jurassic and Cretaceous, in the Tertiary. The erosion was particularly severe near Ashbourne in the southern part of the Peak District where, in places, Triassic sandstones rest directly on an eroded surface of Carboniferous Limestone with the whole of the intervening Upper Carboniferous succession missing.

During post-Carboniferous times the rocks were subjected to moderately strong folding and faulting. The axes of the folds trend in a NNW–SSE direction. The main effect of the folding was to cause the central part of the Peak District to arch upwards and form a broad **anticline** with gently dipping limbs on its eastern and western flanks. Minor folds, such as the Goyt **Syncline** in the west and the Ashover Anticline in the east, display much tighter folding (Fig. 1).

**Faults** are numerous throughout the Peak District but in most cases their **throw** is only a few metres. There is a tendency for the faults to be aligned in either a north-west to south-east or a north-east to south-west direction. The faults subsequently played a major part in easing the passage of hot mineral-bearing solutions through the rocks. The stratified limestones were particularly susceptible to the process of **hydrothermal mineralisation**, probably because they are so well jointed, fissured and, in places, cavernous, as well as dislocated by faults, that the hot solutions were able to percolate freely through the rocks.

As the solutions cooled, various minerals were deposited on the walls of the cavities and fissures to form a network of mineral veins. The thickest and most persistent veins, known as **rakes** to the Derbyshire lead miners, are almost always coincident with faults and it is clear that the presence of fault planes was a major factor controlling the distribution of the mineral veins.

The minerals commonly found in the veins include calcite, **fluorite**, **galena**, **sphalerite** and **barite**. Much the best places for studying these minerals are on the waste tips of abandoned mines situated along the rakes. Many of the larger rakes can be traced for several kilometres across the

**Figure 7** A typical rake near Castleton.

limestone plateau, and the disturbed nature of the land surface caused by the mining operations is a characteristic feature of Peak District topography (Fig. 7).

In most veins there are several minerals present. In some veins the minerals occur in separate layers, but more often than not the crystals of the various minerals are mixed haphazardly together. The calcite, known locally as 'spar', is usually quite easy to recognise. It occurs as large, white crystals with a glassy lustre. It is brittle and can readily be broken along flat **cleavage planes** to produce small crystals, each face of which has the shape of a parallelogram.

The fluorite (also known as fluorspar) crystals are sometimes easy to spot, at other times difficult. Well formed crystals are cubic in shape and, when broken, they cleave across the corners of the cubes. The colour is very variable; white, pale yellow, deep purple and black are the common shades in the Peak District. Frequently the mineral occurs as a mass of small, irregular crystals with a sugary texture.

The most distinctive feature of barite is that for a lightly coloured mineral it is unusually heavy (hence the miners' name for it — 'heavy spar'). It usually occurs as irregularly shaped lumps which, when broken open, can be seen to consist of radiating, fibrous crystals which may be white but more often are coloured in pale shades of yellow, pink or brown.

Galena, the principal ore of lead, is a very heavy mineral. It is easily recognised by its very bright, silvery-grey metallic lustre on freshly exposed surfaces. It can be broken easily along flat cleavage planes which are at right angles to one another. Broken fragments tend, therefore, to have a cubic shape. The silvery lustre of galena tarnishes on exposure to the atmosphere and the mineral then has a dull, blue—grey colour.

Sphalerite, the chief ore of zinc, is usually dark brown in colour. It is a brittle mineral and it cleaves readily in several directions. Freshly exposed surfaces have a dull, glassy, somewhat resinous, lustre.

When copper is present in the veins, as at Ecton, even if only in small amounts, the characteristic bright green stains of **malachite** and **chrysocolla** on weathered surfaces are evident. The principal copper ore, **chalcopyrite**, is a rich golden yellow mineral with a bright metallic lustre.

The final stages in the geological history of the Peak District began with the retreat of the sea which had covered the area during the later stages of the Cretaceous and the subsequent rising of the Pennine uplands during the Tertiary.

In the warm, moist climate of the Tertiary numerous **sinkholes** formed on the limestone plateau in the area between Buxton and Ashbourne. The sinkholes vary in size from small hollows only a few metres across to large depressions more than 60 m deep and 100 m wide. During Pliocene times, towards the end of the Tertiary, the sinkholes became filled with multicoloured sands and clays. The clays from these 'pocket' deposits are used to make fire-bricks and furnace linings.

The Quaternary Period is represented by thin but widely distributed sediments of Pleistocene and Holocene (Recent) age. Patches of **boulder clay** containing striated erratic boulders of local and northern origin occur on the limestone plateau. They indicate that the area was, on at least one occasion, overridden by an ice sheet moving down from the north.

Deposits of **head**, consisting of sand and clay containing angular blocks of sandstone up to a metre or so across, are common on slopes below the gritstone escarpments in Edale and the Kinder Scout area. They are thought to have been formed by the weathering of the escarpments during the intensely cold episodes of late Pleistocene time.

During the Holocene extensive deposits of peat accumulated on the gritstone plateaux, as at Kinder Scout, while in the steep-sided valleys landslips were a common feature. Some of the landslips are still active from time to time, especially the one on the eastern face of Mam Tor (Fig. 17).

The location of the excursions described in Part 2 is shown in Figure 8.

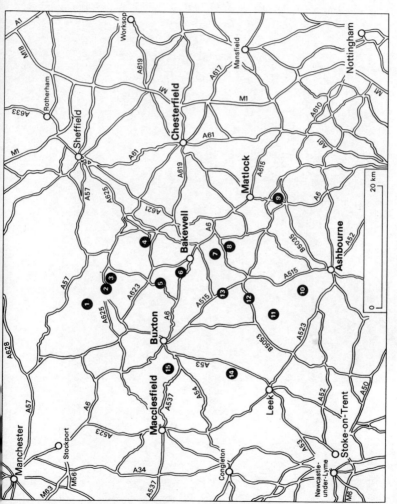

**Figure 8** Map showing the location of excursions 1–15.

# Part 2: Field excursions

# 1 Edale and Kinder Scout

(1 day)

*Purpose*: On this excursion the Namurian succession from the Edale Shales up to and including the Kinderscout Grit is examined.

The geological structure of the area is simple. All the beds dip at low angles and are little disturbed by faults. The route, along minor roads and tracks, involves a climb of about 300 m up from the Edale Valley to the plateau of Kinder Scout, crossing part of the plateau and descending back into the valley. The total walking distance is either 8 km or 11 km depending upon which of two alternative routes is taken. Boots and appropriate clothing should be worn, and a compass carried.

On certain weekdays during the grouse shooting season in late August and September some parts of the moors may be closed to the general public. Notices giving details of the closures are posted at the access points to the moors and at the National Park Information Centre in Edale (telephone no. Edale 207).

*OS maps*:    1:50 000 Sheet 110
              1:25 000 The Dark Peak Outdoor Leisure Map
*IGS map*:    1:50 000 Sheet 99 Chapel-en-le-Frith (S or D)

The excursion starts at Edale Station. At present there is a good service of trains to Edale from Manchester and Sheffield. Close to the station there is a large free car park with toilet facilities (point P on Fig. 9). Cafés and hotels in Edale village provide refreshments and accommodation. There are several camp-sites in the area and at Nether Booth, about 2 km east of Edale, there is a large Youth Hostel.

**Succession**

|  | Thickness (*m*) | Zones |
|---|---|---|
| Kinderscout Grit | more than 60 | |
| Shales and thin sandstones | 80 | |
| Shale Grit | 100 | $R_1$ |
| Mam Tor Beds | 70 | |
| Edale Shales | more than 100 | $E_2 - R_1$ |

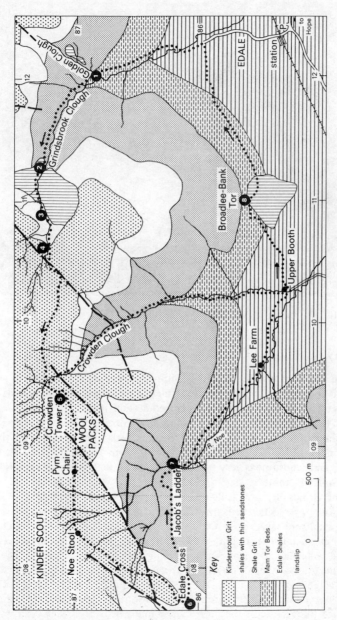

**Figure 9** Map of the Edale and Kinder Scout area.

Key

Kinderscout Grit

shales with thin sandstones

Shale Grit

Mam Tor Beds

Edale Shales

landslip

0    500 m

KINDER SCOUT

Noe Stool

Pym Chair

Crowden Tower

WOOL PACKS

Edale Cross

Jacob's Ladder

Crowden Clough

Grindsbrook Clough

Golden Clough

Broadlee-Bank Tor

Upper Booth

Lee Farm

R. Noe

EDALE

station

to Hope

**Figure 10**   The start of the Pennine Way at Edale.

## Itinerary

Proceed north through Edale village, join the Pennine Way footpath and follow it up the Grindsbrook valley. Straight ahead there are splendid views of the gritstone escarpment of Kinder Scout (Fig. 10).

At **1** (121868), where the main footpath crosses the tributary stream descending from Golden Clough, grey shale beds interspersed with thin sandstones are exposed. These are part of the Mam Tor Beds. The mixture of shale and sandstone is very typical of much of the Namurian succession in the South Pennines.

As the Grinds Brook is followed further up the valley exposures of the Shale Grit appear. Shale beds now form only a minor part of the succession but the sandstone layers are much thicker than in the Mam Tor Beds. They are best exposed in the stream bed where the Grinds Brook cascades over a series of waterfalls, each one formed by a near horizontal layer of hard, fairly fine-grained sandstone with partings of micaceous silt and shale.

At **2** (113873) the path crosses a tract where landslips are occurring. The numerous repairs to, and diversions of, the footpath hereabouts are usually attributed to over-use of the path by too many walkers, but in reality the slope of the hillside is inherently unstable so that the slow downhill creep is certain to continue whether people walk over it or not. Large boulders of coarse-grained pebbly sandstone are prominent in the landslip area. These are from the Kinderscout Grit escarpment and they contrast strongly with the much finer sandstone of the Shale Grit outcrops in the stream bed.

The gradient of the footpath eases temporarily after the crossing of the landslipped area. The change is due to a reversion to a predominantly soft shale succession. At **3** (109873) the dark grey micaceous shales are well exposed in a section stretching along the stream bed for about 100 m. An interesting system of regularly spaced joints is well displayed by some of the thin beds of sandstone that accompany the shales (Fig. 11).

**Figure 11** Chemical weathering along joints in sandstone in Grindsbrook Clough (× ½).

Where the Grinds Brook divides take the left fork and ascend the now very steep and bouldery path (it is also the stream bed). The hard going here provides an ample opportunity to observe the Kinderscout Grit at close quarters so that it becomes impressed on the mind as well as the feet (Fig. 12). At the top of the section the stream runs over solid rock and erodes potholes in the coarse sandstone. From this point (**4**) (107873) there is a fine view back down Grindsbrook Clough beginning with the steep scarp of the Kinderscout Grit rising above the bench formed by the shales and thin sandstones. Below the bench there is a less prominent scarp

**Figure 12**   Outcrops of the Kinderscout Grit in Grindsbrook Clough.

formed by the Shale Grit and Mam Tor Beds which in turn levels off on the Edale Shales at the bottom of the valley.

From **4** the route lies westwards towards Crowden Tower crossing a wide peat-covered plateau on the way. The peat layer, which is generally 2–3 m thick, has in places been so severely eroded by the strong winds which frequently sweep across the plateau that deep gullies have been formed in it. In some of the gullies the peat has been completely stripped away, revealing the gritstone bedrock. The stumps of quite large trees can sometimes be seen embedded in the lower layers of the peat exposed in the sides of the gullies. Pollen analyses of these lower layers indicate that deposition of the peat commenced about 7000 years ago towards the end of the Boreal climatic phase of the Flandrian Stage of the Pleistocene.

The footpath is not at all clearly defined across the peat-covered plateau. In misty weather great care should be taken on this part of the route. It is only too easy to lose all sense of direction and become lost.

The outcrops of Kinderscout Grit on and around Crowden Tower (**5**) (095871) have been carved into extraordinary shapes as a result of centuries of exposure to strong winds and heavy rain. Any weaknesses in the rocks have been exploited so that joints have been enlarged and softer layers of rock etched out (Fig. 13).

**Figure 13**  The Kinderscout Grit near Crowden Tower.

From Crowden Tower there is a choice of two routes back to Edale. The shorter route follows the steep footpath down Crowden Clough to Upper Booth. In rapid order the outcrops of the Kinderscout Grit, shales and thin sandstones, Shale Grit and Mam Tor Beds are crossed before the gradient eases on the outcrop of the Edale Shales. These are exposed at several

**Figure 14**  The landslip at Broadlee-Bank Tor, Edale.

places near Upper Booth where the meandering stream is cutting into its steep banks.

The longer route takes a westerly track from Crowden Tower threading a way through a maze of weathered blocks of gritstone at the Wool Packs and crossing the tors of Pym Chair and Noe Stool before descending gradually to meet the ancient pack-horse road from Hayfield to Edale at Edale Cross (6) (077861). From this point, on the main watershed of the Pennines, drainage to the west is into the Mersey Basin and the Irish Sea, while to the east the Trent system drains into the North Sea.

The strata in the area around Edale Cross are displaced by several minor faults but the fault planes are not exposed. Scattered small outcrops of the Shale Grit appear beside the footpath down Jacob's Ladder and in the stream bed at the foot of the steep slope grey shales and thin sandstones of the Mam Tor Beds are seen (7) (088862). These continue to be exposed at intervals in the banks of the River Noe for the next 600 m or so until replaced by the underlying Edale Shales. These, too, are well exposed and consist of dark grey shales with scattered bands of ironstone nodules. Fossils are not abundant in the Edale Shales, but flattened specimens of the goniatites *Reticuloceras* and *Homoceras* and the bivalves *Posidonia* and *Dunbarella* can be found at some of the outcrops.

From Upper Booth take the footpath through the fields to Broadlee-Bank Tor (8) (110857). Here there is a perfect example of a small-scale landslip. The slip was probably caused by the pressure of groundwater in the Mam Tor Beds. This allowed the soft shales to lose much of their bearing strength so that they were squeezed out by the weight of the overlying Shale Grit sandstones. Both the Mam Tor beds and the Shale Grit are exposed in the face of the landslip, while the toe of the slip, with its distinctive irregular topography, stretches down the hillside for about 300 m (Fig. 14).

From 8 the footpath leads directly back to Edale village.

# 2 Castleton Area: Treak Cliff and Mam Tor

(1 day)

*Purpose*: On this excursion reef limestones in the Carboniferous Limestone (Dinantian) Series, and shales and sandstones in the Millstone Grit (Namurian) Series are examined.

The route is along public footpaths and minor roads. The walking distance is about 7 km. In wet weather the steeper sections of the route can be very slippery and appropriate footwear is advisable. Safety helmets should be worn when examining the outcrops at the foot of Mam Tor.

| *OS maps*: | 1:50 000 Sheet 110 |
| | 1:25 000 The Dark Peak Outdoor Leisure Map |
| *IGS maps*: | 1:50 000 Sheet 99, Chapel-en-le-Frith (S or D) |
| | 1:25 000 Sheet SK18, Edale & Castleton |

The excursion starts at the public car and coach park in Castleton (149830, point P on Fig. 15). There are toilet facilities at the car park. Several cafés and hotels in the village offer refreshments, meals and accommodation. Other amenities include a Youth Hostel and a National Park Information Centre.

## Succession

| | | Thickness (*m*) | Zones |
|---|---|---|---|
| Namurian | Mam Tor Beds | about 150 | $R_1$ |
| | Edale Shales | about 130 | $E-R_1$ |
| | unconformity ~~~~~~~ | | |
| | Eyam Shales | 0–15 | $P_2$ |
| | Beach Beds | 0–10 | $P_2$ |
| Viséan | unconformity ~~~~~~~ | | |
| | Bee Low Limestones and reef complex | 100 | $D_1$ |

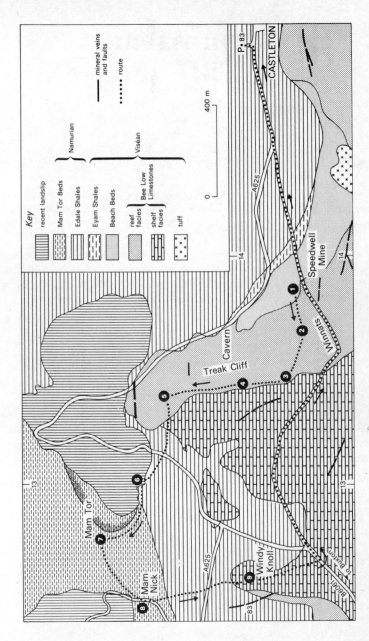

**Figure 15** Map of the Castleton, Treak Cliff and Mam Tor area.

**Itinerary**

On leaving the main car park in Castleton turn right and proceed westwards along the main A625 road for about 300 m, then take the minor road on the left leading to the Speedwell Cavern (139828). The deep, narrow, twisting cleft of the Winnats lies directly ahead. It merits a close study. Theories which have been proposed for the origin of the gorge include: (a) severe erosion during the final stages of the last glaciation by a river fed from the melting of the ice sheet which covered much of northern England; (b) collapse of the roof of an extensive underground cave system; (c) scouring by submarine currents active during the time of deposition of the reef complex limestones, thus forming a surge channel similar to those found in modern coral reefs; and (d) various combinations of the above theories. As yet there is no agreement as to which, if any, of these is the correct explanation of the origin of the gorge.

Behind the car park opposite to the entrance to the Speedwell Cavern there is a small quarry in the Beach Beds (**1**) (138828). In the quarry are well bedded limestones composed very largely of fragments of the shells of large brachiopods. Many, but not all, of the fragments are very well rounded. Originally the rock must have been a coarse shell-sand, and one theory of its origin is that the sand was deposited in a turbulent environment formed by waves breaking on a beach, hence the name of the formation: the Beach Beds. An alternative explanation is that if the Winnats gorge was in existence as a submarine surge channel at the time of deposition of the Beach Beds, strong tidal currents in the channel might have transported the shell fragments to their present position.

Evidence from boreholes sunk near Castleton shows that the Beach Beds are partially interbedded with Eyam Shales containing fossils of known $P_2$ age. Immediately underneath the Beach Beds come reef limestones containing $D_1$ fossils. The whole of zone $D_2$, which in other parts of the Peak District consists of limestones up to 200 m thick, is missing in the Castleton area. There is, however, no visible evidence of unconformity here because both the Beach Beds and the reef limestones are dipping in the same direction and at much the same angle.

From **1** ascend for about 150 m along the ridge which separates the Winnats from Treak Cliff. Location **2** (137828) is a small quarry beside the only tree growing on this part of the hillside. Fossils are abundant here. Shells of productid brachiopods and the bivalve *Pseudamussium* predominate. The radiating brown marks visible on the outer surfaces of some of the bivalve shells are thought to be traces of their original colour.

From **2** continue upwards along the ridge for another 200 m or so to the entrance into the Old Tor Mine at **3** (135828). The mine has long been abandoned and has little of interest inside, but at the upper entrance the reef limestone is cut by numerous thin veins and patches of the banded, purple and white variety of fluorite known locally as Blue John. Please do not attempt to deface this exposure. The minerals belong to the National Trust and are there to be admired, not to be removed.

From **3** head northwards, crossing the fence at a stile and proceed to **4** (134830), the highest point on Treak Cliff. Here the limestone is unusually pale in colour, very fine-grained and poorly bedded. A close inspection of the weathered surfaces reveals the characteristic concentric layering of algal growths. The branching coral *Lithostrotion* is abundant here. Some of the colonies exceed 1 m in diameter. The algal and coral limestones can be traced along the crest of Treak Cliff for about 150 m before they apparently die out.

At **5** (134834), a craggy outcrop to the north of the footpath leading to Treak Cliff cavern, reef limestone is very well exposed. The rock is very rich in fossils. Brachiopods and corals occur abundantly. Crinoids, bryozoa, gastropods, goniatites and fragments of trilobites can also be found.

Much can be deduced from studying cross-sections of the brachiopod shells embedded in the limestone. Some of the shells are completely hollow

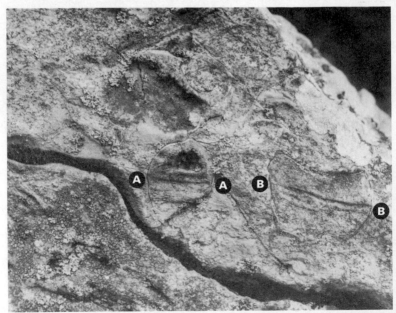

**Figure 16** Geopetal infills in brachiopod shells (A–A and B–B) on Treak Cliff (×1).

and empty; others are either partially or fully filled with limestone similar to that outside the fossils (Fig. 16). The fillings, known as geopetal infills, were formed by the percolation of lime-rich mud and sand into hollow, empty shells as they were being buried in the sediments accumulating on the reef. At the time of burial the bedding layers and upper surface of the infilling sand and mud would have been horizontal; but measurements of the present dip of the layers *inside* the shells now give an average reading of 8° to the north-east, indicating that the Treak Cliff area has been tilted in that direction at some time *after* the limestones had become lithified. The dip of the bedding planes in the limestone *outside* the shells is, however, much higher and averages 35° in the same northeasterly direction. From these figures it is argued that the present dip of the reef limestone is a combination of 27°, representing the original slope of the surface of the reef while it was being formed, and 8° derived from the later tilting of the area as a result of tectonic movements.

Treak Cliff Cavern, near **5**, is open to the public every day of the year except Christmas Day and is well worth a visit. Fluorite mineralisation of the limestone is exceptionally well displayed in the cavern and there is a fine natural cave with many magnificent stalactites and stalagmites on show. Hot and cold drinks and light refreshments are on sale at the cavern and there are covered and outdoor picnic facilities where, if you wish, your own food may be eaten. Parties wishing to visit the cavern may book in advance

**Figure 17** The east face of Mam Tor.

by telephoning Hope Valley (0433) 20571. Cars and coaches can be parked at the side of the A625 road close to the entrance to the cavern.

From Treak Cliff follow the footpath to Mam Tor. At 517 m the summit is the highest point in the area and the sheer eastern face of the mountain forms a dramatic and conspicuous landmark as well as providing one of the best exposures of Namurian strata in the district (Fig. 17). The exposure owes its origin to a great landslip in geologically recent times. The extensive area of hummocky, badly drained ground to the east of Mam Tor is another result of the landslip. Movements are still taking place from time to time and have caused widespread damage to the main road. The reason for the unstable nature of the hillside can be seen by examining the exposures at **6** (130835). The rock is a soft, crumbly black shale which, when wet, rapidly reverts to its original muddy state and has little bearing strength.

The black shale is part of the Edale Shales at the base of the Millstone Grit Series (Namurian). Within the shale beds hard calcareous concretions, or bullions, up to 0.5 m in diameter occur. When broken the bullions are frequently found to contain well preserved fossils of goniatites and bivalves. These signify a marine origin for the shales and help to place them in Zones E to $R_1$ of the zonal succession.

The slopes at the foot of the east face of Mam Tor are littered with numerous blocks of brown, micaceous sandstone which have tumbled

**Figure 18**   Sole marks (flute casts) on sandstone at Mam Tor.

down from the higher parts of the cliff (the wearing of safety helmets is advisable here). Many of the blocks, which are from the Mam Tor Beds, show well defined **sole marks**, including **flute marks**, **groove marks** and **load casts** (Fig. 18). These were formed by the deposition of sand on top of an irregularly grooved, rippled and channelled surface of a layer of mud or silt. Later, under the weight of the overlying sediments, the soft layers of sand, silt and mud were compacted and somewhat distorted. The resulting sole marks preserved on the base of the sandstone beds thus form a negative imprint of the surface of the underlying sedimentary rocks.

The alternating layers of sandstone and mudstone in the Mam Tor Beds are thought to indicate deposition in a **turbidite** environment, possibly at the seaward end of a river delta. Fossils are few and consist mainly of fragments of plants.

A path leads up the south side of the Mam Tor landslip to **7** at the summit (128836). From here there are extensive views across the Edale Valley to Kinder Scout in the north and across the limestone plateau to the south.

From **7** follow the footpath down to **8** at Mam Nick (124834). Here, in the roadside cutting, black mudstones and thin sandstones typical of the Mam Tor Beds are exposed. The beds are cut by a small normal fault with a throw of about 2 m (Fig. 19).

Figure 19   A small fault in the Edale Shales at Mam Nick.

The hillside on the north-west side of Mam Nick has suffered extensive landslipping and, although now well grassed over, the landslipped area is easily recognisable by its very irregular and knobbly topography.

From Mam Nick take the footpath southwards to the limestone quarry at Windy Knoll (9) (126830). This footpath is part of the ancient Old Portway track which ran from Nottingham up through the Peak District and dates back more than 3000 years to the late Neolithic or early Bronze Age.

There are two interesting phenomena to examine at Windy Knoll. One is the occurrence of fissures which were cut down into the reef limestones, perhaps as a consequence of a period of **karstic weathering**, and later were filled with a mixture of angular boulders of light grey limestone embedded in a matrix of dark grey, muddy limestone. These features are known as 'Neptunean Dykes'.

The second feature of note is the impregnation of the upper layers of limestone in the quarry with elaterite, a natural bitumen. It is a dark brown, sticky, rubbery substance with a strong oily smell. The occurrence of elaterite is very rare in Britain.

From Windy Knoll return to Castleton via the footpath southwards across the fields and down the Winnats.

# 3    Castleton Area: Cave Dale and Dirtlow Rake

(1 day)

*Purpose*: Various facies of the Carboniferous Limestone (Dinantian) Series are examined on this excursion. In Cave Dale marginal reef limestones are seen to pass laterally into shelf limestones. Interbedded with the shelf limestones is an extrusion of basalt. Additionally there are several veins bearing calcite, fluorite, barite and galena. Along Dirtlow Rake there are extensive mine workings in veins of this type.

The route follows public footpaths, tracks and minor roads. The walking distance is about 6 km. There are many abandoned mine shafts in the area. Warning notices are posted round most of these and on no account should they be ignored as most of the shafts are very deep and in a dangerous condition.

*OS maps*:    1:50 000 Sheet 110
                   1:25 000 the Dark Peak Outdoor Leisure Map
*IGS maps*:   1:50 000 Sheet 99, Chapel-en-le-Frith (S or D)
                   1:25 000 Sheet SK18, Edale & Castleton

The excursion starts at the free car and coach park in Castleton (149830, point P on Fig. 20). There is a frequent bus service to Castleton from Sheffield via Hathersage and Hope. There are toilets at the car park. In the village there are several cafés and hotels, a Youth Hostel and a National Park Information Centre.

## Succession

|  | | Thickness (*m*) | Zones |
|---|---|---|---|
| Viséan | Monsal Dale Limestones (shelf facies) | 100 | $D_2$ |
| | Bee Low Limestones (shelf and reef facies) | 100 | $D_1$ |

## Itinerary

The entrance to Cave Dale is situated at the southeastern corner of the

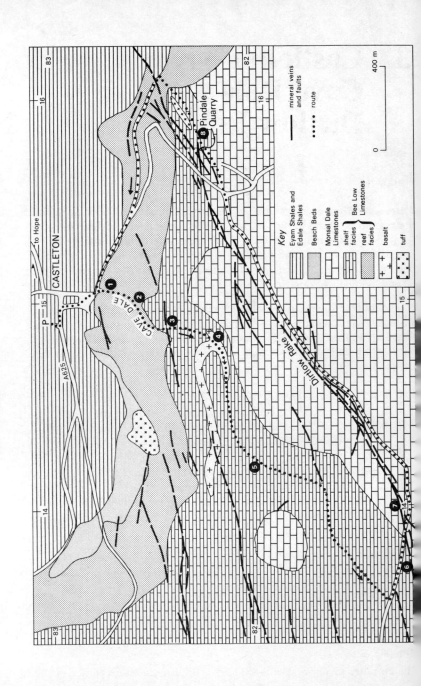

Market Place in Castleton. The way into the dale is signposted through a narrow gorge in the reef limestones. These can readily be examined in the small, disused quarry at **1** (151827). The poorly stratified beds dip northwards at about 20° and contain a few fragmented crinoids and brachiopods. Two thin veins of calcite with specks of galena run vertically up the face of the quarry.

Patches of limestone thickly crowded with such fossils as corals, brachiopods or goniatites occur in the reef. Normally in any one patch one type of fossil will predominate in numbers over all the others. A good example of such an occurrence is seen at **2** (150826), a craggy outcrop of limestone on the southern side of the footpath. Small brachiopods belonging to the genus *Pugnax* occur in great abundance near the foot of the crag, but few other fossils are found here.

At **3** (148824) a vein of coarsely crystalline calcite is seen on both sides of the dale. The vein is about 2 m thick and at the margins of the vein thin

**Figure 21** A mineral vein in Cave Dale.

streaks and patches of galena and purple fluorite occur among the crystals of calcite (Fig. 21).

The limestones in the vicinity of **3** differ from those seen lower down the dale in that the stratification is more clearly defined and the bedding planes are nearly horizontal. These changes are indicative of a transition from the reef facies to the shelf facies.

From **3** continue along the footpath up the dale. Soon the gradient steepens and the path deteriorates, becoming rough, wet and muddy. The reason for the steeper gradient and the very damp conditions underfoot is the occurrence of a bed of basalt within the limestone succession. The basalt is well exposed at **4** (148822).

In the lower part of the outcrop dark grey rock of medium grain-size is seen. It has a roughly columnar joint structure (Fig. 22). This rock was

**Figure 22** An outcrop of basalt in Cave Dale.

originally an olivine-basalt, but as a result of extensive chemical weathering many of the feldspar and ferromagnesium minerals have been replaced by clay minerals, chlorite and calcium carbonate. The rock in the upper part of the outcrop is vesicular and some of the vesicles contain amygdales of calcite and chlorite.

The presence of vesicles suggests that the basalt originated as a sub-aerial lava flow which was extruded over the area during a temporary recession of the shallow sea which previously had covered the area while the limestones were being deposited. Scattered outcrops of vesicular basalt and patches of rust-brown soil presumed to be derived from the weathering of basalt extend across the fields for a distance of about 800 m to the west of **4**. What appears to be the same bed of basalt is exposed again in the Buxton area and in Miller's Dale and Tideswell Dale (see excursion 5), where it is known as the Lower Miller's Dale Lava.

At the foot of the dip slope of the reef complex at 144825 there is a poorly exposed outcrop of tuff which was at one time thought to be part of a volcanic neck penetrating through the reef limestones as an intrusion. It is therefore referred to as the Speedwell Vent. An alternative explanation, for which there is now some supporting evidence, is that the tuff is part of the same lava flow which produced the basalt in Cave Dale but, unlike the basalt, it was cooled rapidly under water with consequent fracturing and disintegration of the rock. The precise location of the vent from which the basalt and, possibly, the tuff were extruded is not known but presumably it lies some distance away to the south of Cave Dale.

From **4** continue along the footpath up the dale. This is now a shallow, open valley with gently sloping sides, in marked contrast to the deep, steep-sided gorge at its lower end. Outcrops of nearly horizontal beds of Bee Low Limestones occur here and there. Generally fossils are not too plentiful in these exposures but at **5** (143820) there are several colonies of the compound coral *Lithostrotion*.

Follow the footpath until it meets a minor road at 135813, then turn left. The road runs parallel to Dirtlow Rake, one of the major mineral veins of the area. Several important lead mines operated here during the nineteenth century. The first of these, the Hazard Mine, is on the right of the road at **6** (137813). The shaft is about 200 m deep and was connected at depth with the Hollandtwine Mine 300 m further along the road at **7** (140812). The waste material from these mines included much calcite, fluorite and barite. In recent years the tips have been extensively reworked for the fluorite which is now in much greater demand than it used to be.

It is dangerous to enter any of the mined areas but there is still plenty of waste lying at or near to the roadside. Specimens of cleaved calcite are particularly abundant here along with the cockscomb variety of barite. In contrast to the white calcite the barite is normally either pink or pale brown in colour. Specks of galena are common in the barite, as are patches of purple fluorite.

Dirtlow Rake is a composite vein with several branches which divide and unite repeatedly. As well as being deep-mined the veins have been worked in a series of open cuts all the way along the rake to Pindale Quarry. As a result of this the ground adjacent to the road is much disturbed.

From the road junction at 155821 take the footpath down Pin Dale to **8**

**Figure 23**   Pindale Quarry (lower level) and Hope cement works.

(159823). Limestone was formerly quarried here on a large scale on two levels (Fig. 23). Grey crinoidal limestone of the Monsal Dale Beds is seen in the upper level. Nodules of chert occur in some of the beds. Calcite veins and patches of fluorite in both purple and pale yellow varieties are also present.

In the lower level of the quarry there are massive beds of the Bee Low Limestones, some of which are rich in large crinoids and the brachiopod *Gigantoproductus*. Black chert also occurs in some of the beds. The layer of tuff which underlies the limestones of the lower level is believed to be of the same age as the basalt in Cave Dale.

From Pin Dale return to Castleton along the minor road skirting the base of the frontal slope of the reef.

# 4 Eyam, High Rake and Coombs Dale

(1 day)

*Purpose*: The industrial exploitation of geological resources is the principal feature demonstrated on this excursion. Limestone quarrying is a major industry in this area along with the mining and processing of fluorite. Carboniferous limestone is quarried for roadstone, and the fluorite obtained from several veins cutting through the limestone is used in the manufacture of steel and many chemical compounds.

The walking distance on this excursion is about 8 km over public footpaths, tracks and minor roads. The working quarries and mines are private property and should not be entered, but can easily be viewed from adjacent footpaths.

*OS maps*:    1:50 000 Sheet 119
                1:25 000 The White Peak Outdoor Leisure Map
*IGS maps*:  1:50 000 Sheets 99, Chapel-en-le-Frith (S or D) and 111, Buxton (S or D)

The excursion starts at Eyam (220764), a pleasant village situated on the margin between the outcrops of Viséan limestones and Namurian shales. There is a large, free car and coach park with toilets in the village, and cafés and hotels for refreshments. There are Youth Hostels at Beech Hurst, north of the village and at Bretton, 3 km to the north-west. Buses on the Sheffield–Buxton route stop at Eyam.

## Succession

|  |  | Thickness ($m$) | Zones |
|---|---|---|---|
| Namurian | shales and mudstones | 200 | E |
| ~~~~~ unconformity ~~~~~ |  |  |  |
| Viséan | { Eyam Limestones | 40 | $P_2$ |
|  | { Monsal Dale Limestones | 150 | $D_2$ |

The Eyam Limestones are thinly bedded and generally dark in colour. The Monsal Dale Limestones consist of more thickly bedded limestones which are, in the main, pale in colour. In both formations patches of poorly bedded reef limestone also occur.

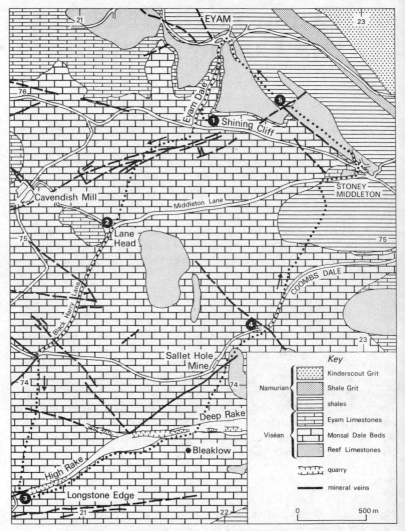

**Figure 24** Map of the Eyam, High Rake and Coombs Dale areas.

## Itinerary

From the square in the centre of Eyam proceed southwards down the A6010 road through the wooded valley of Eyam Dale for about 500 m. At the electricity substation, just before reaching the junction with the A623 road, turn left along a footpath which leads to Shining Cliff (1) (219758).

Here a very fine section of the uppermost strata of the Monsal Dale Limestones and some of the overlying Eyam Limestones is exposed (Fig. 25).

**Figure 25** Shining Cliff. The band of pale limestone about two-thirds of the way up the cliff is the White Bed.

| | | | Thickness (*m*) |
|---|---|---|---|
| Eyam Limestones | 1. | Dark limestone with chert nodules | 10 |
| | 2. | Massive pale oolitic limestone | 2 |
| | 3. | Pale cherty limestone | 12 |
| | 4. | *Dibunophyllum* Bed – grey limestone with *Dibunophyllum* and *Lithostrotion* | 0.3 |
| | 5. | Pale cherty limestone | 1 |
| | 6. | The White Bed – very pale, fine-grained cherty limestone | 2 |
| Monsal Dale Limestones | 7. | Pale grey cherty limestone, shelly in places | 9 |
| | 8. | The Upper Shell Bed – pale limestone rich in *Gigantoproductus* | 0.5 |
| | 9. | Massive pale limestone with compound corals and *Gigantoproductus* | 7 |
| | 10. | The Black Bed – very dark grey limestone with *Lonsdaleia* | 1.5 |
| | 11. | Pale grey limestone with abundant corals | 1.5 |

For much of the way the footpath is on the upper surface of the Black

Bed, and the prominent bumps on this surface are colonies of the compound coral *Lonsdaleia*.

In the lower part of the cliff, below the footpath, there are further exposures of pale grey limestones some of which are shelly.

**Figure 26**   The industrial side of the Peak District – Darlton Quarry, near Stoney Middleton.

From **1** return to Eyam Dale, turn left along the A6010, then cross the busy A623 road and take the clearly signposted bridle-path which passes through the limestone quarries on the south side of the road (Fig. 26). The quarries, in the Monsal Dale Limestones, are being worked for roadstone and the working faces are clearly visible from the footpath. From a point on the path near the top of the quarries there is a fine view back towards Shining Cliff. The White Bed is the most conspicuous feature. This distinctive marker band can be traced along both sides of Middleton Dale for more than a mile.

At Lane Head (**2**) (212751), where the bridle-path crosses Middleton Lane, *Lithostrotion*-bearing limestone high in the Monsal Dale Limestones is seen in a small roadside quarry.

Continue south along the bridle-path, now known as Black Harry Lane after a notorious highwayman who frequented these parts in the early eighteenth century and was eventually hanged at nearby Wardlow Mires for his misdeeds. The path skirts the large settling ponds of the big fluorite-processing plant at Cavendish Mill.

At **3** (206732) the path reaches the High Rake, the site of one of the largest opencut fluorite mines in Britain. Although mining has now ceased on the site and parts of it are being backfilled with mine waste there is still much of interest to be seen here. Good specimens of yellow fluorite, white calcite, pink barite and silvery-grey galena are not hard to find.

The road from **3** follows the track eastwards along the crest of Longstone Edge. On the right there are spectacular views of the countryside to the south and, on the left, the fluorite mines along the rake (Fig. 27).

**Figure 27**   High Rake on Longstone Edge.

Turn left opposite the entrance to Bleaklow Farm and follow the footpath leading down into Coombs Dale. The path narrows as it descends diagonally down the steep scree-covered slope of the dale, so take care. The path overlooks the entrance to the Sallet Hole mine where, by means of an adit driven into the hillside, fluorite is obtained at depth from the Deep Rake. Some old mine workings, now grassed over, are crossed by the path on the way down and there are also a few scattered outcrops of shelly,

**Figure 28**    *Lithostrotion* in Coombs Dale ( × ½).

crinoidal limestone, but the most interesting exposures are on the crags on the north side of the mine road at the bottom of the dale at **4** (222744).

The crags contain an abundance of corals, including *Caninia*, *Dibunophyllum* and *Lithostrotion*, and, as a result of prolonged exposure to the weather, the fossils stand out conspicuously from the more soluble limestone in which they are embedded (Fig. 28). This coral bed is probably at the same stratigraphical horizon as the Hobs House Coral Bed in Monsal Dale (see excursion 6).

Follow the main road down Coombs Dale for about 200 m then turn left on to an unsignposted footpath which climbs out of the dale and through the fields to Stoney Middleton. A second path branches off at a point another 500 m or so down the road, should the first one have been missed.

Cross the A623 at the Royal Oak in Stoney Middleton, turn right at the old toll-house and proceed up the Bank for a short distance, noting outcrops of limestone full of *Gigantoproductus* at the roadside. Take a sharp turn at first left into Cliff Bottom and Mill Lane, then follow the clearly signposted footpath across the fields back to Eyam. This path crosses the dip-slope of a reef in the Eyam Limestones. There were several lead mines in this area at one time and traces of the old workings are obvious. The path passes the edge of the waste tips of the long-abandoned Cliffstile Mine (**5**) (224759). There is much fluorite and calcite here. The limestone blocks in the wall beside the path are rich in brachiopods and crinoids. These should be admired but not removed.

# 5 Tideswell Dale and Miller's Dale

(½ day)

*Purpose*: On this excursion a dolerite sill, two basaltic lava flows and several limestones of the shelf facies of the Carboniferous Limestone (Dinantian) Series are seen.

The route is about 5 km long and, for most of the way, follows public footpaths and minor roads.

*OS maps*:     1:50 000 Sheet 119
              1:25 000 The White Peak Outdoor Leisure Map
*IGS maps*:    1:50 000 Sheet 111, Buxton (S or D)
              1:25 000 Sheet SK17, Miller's Dale (S & D)

The excursion can be started either at the Tideswell Dale picnic site (154742), where there are toilets and ample parking space, or in the village at Miller's Dale (142733) where parking space is limited but refreshments are available.

There are bus services to Tideswell Dale and Miller's Dale from Buxton, Bakewell and Sheffield. At Ravenstor there is a Youth Hostel.

## Succession

| | | Thickness (*m*) | Zones |
|---|---|---|---|
| | Monsal Dale Limestones | over 100 | |
| | Upper Miller's Dale lava (basalt) | 25 | $D_2$ |
| | Monsal Dale Limestones | 0–10 | |
| Viséan | Miller's Dale Limestones | 60 | |
| | Lower Miller's Dale Lava (basalt) | 20 | $D_1$ |
| | Dolerite sill intrusion | 0–25 | |
| | Chee Tor Rock (limestone) | 80 | |

## Itinerary

If the start is made at the Tideswell Dale picnic site, the first exposures seen are of the dolerite sill. This was formerly quarried at **1** (154742) on the

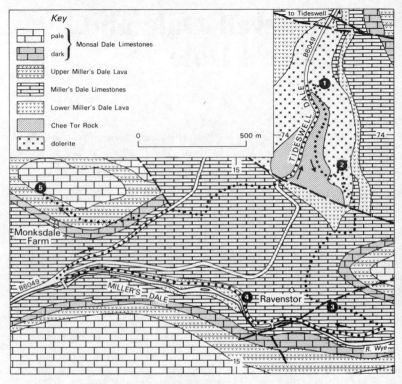

**Figure 29**   Map of the Tideswell Dale and Miller's Dale area.

eastern side of the car park. The dolerite has been subjected to prolonged chemical weathering, particularly along the joint planes where percolating groundwater has hastened the hydration and oxidation of the original feldspar and ferromagnesian minerals. Towards the top of the exposure the rock has disintegrated almost completely to form a rich rust-coloured clay soil in which a few isolated, rounded remnants of the original dolerite are embedded (Fig. 30). The dolerite is interpreted as being part of a sill intruded between the Chee Tor Rock and the Lower Miller's Dale Lava.

From **1** proceed southwards along the footpath down Tideswell Dale noting outcrops of the Chee Tor Rock on the left side of the path. Where the path divides take the left fork up to the main picnic area (**2**) (155738). This, too, was once a working quarry in the dolerite sill and here, also, the dolerite has suffered extensive chemical weathering.

Follow the footpath through the picnic site and back down into the dale to rejoin the main footpath. A fault with a downthrow of about 15 m to the south truncates the dolerite sill outcrop and brings the overlying Miller's

**Figure 30** Weathered dolerite sill in Tideswell Dale. At the top of the sill only the cores of the blocks of dolerite remain.

Dale Limestones down to the dale bottom where they appear as outcrops of well bedded limestone dipping down the dale at a low angle.

At **3** (154732), where the dale begins to bend round towards the east, old mine workings are seen on the left. A vertical vein of calcite about 0.5 m wide is exposed on the cliff face and a few metres further down the dale an adit has been driven into the limestone, presumably to intersect the vein.

Continue down the footpath to its end then turn to the right along the picturesque riverside road up Miller's Dale. At **4** (150733), the spectacular crag of Ravenstor, thick beds of the Miller's Dale Limestones overlie the Lower Miller's Dale Lava (Fig. 31). The junction between the two formations is worth a close examination. The top of the lava, exposed at the base of the cliff, consists of vesicular ashy basalt, reddened in places. It

**Figure 31**  Miller's Dale Limestone at Ravenstor. At the base of the cliff the Lower Miller's Dale Lava is exposed.

is overlain by about 30 cm of soft, greenish-grey chloritic clay followed by the limestone. The soft clay has been preferentially eroded from the cliff, thus providing an excellent exposure of the undersurface of the basal layer of limestone. This surface is very irregular and it would appear that the top of the lava flow underwent considerable weathering, to produce the chloritic clay, and erosion, to produce the irregular surface, before the basal limestone was deposited on it (Fig. 32).

**Figure 32** The eroded junction between the Miller's Dale Limestone (above) and the Lower Miller's Dale Lava (below) at Ravenstor.

From **4** continue up Miller's Dale. After passing (or pausing at) the Angler's Rest Inn, turn sharply to the right, cross the B6049 road and go up Meadow Lane which branches off to the left and is signed 'Unsuitable for motors'. The lane climbs steeply past outcrops of the Miller's Dale Limestones. About 100 m after passing the private entrance to Monksdale Farm cross into the field on the left of the lane at a stile set in the stone wall. A public footpath runs from the stile to the top right-hand corner of the field. About halfway across the field turn left on to another path which leads towards the top left-hand corner of the field. Cross the wooden stile at the top left-hand corner and join a track which runs obliquely up the hillside to **5** (141737). Small outcrops of highly weathered vesicular basalt appear by the side of the track. They are part of the Upper Miller's Dale Lava. Monsal Dale Limestones overlie the basalt and form the crest of the hill. These limestones were formerly quarried on a very large scale on the south side of Miller's Dale, as can be seen from the excellent viewpoint at **5**.

Return to Meadow Lane. About 200 m further up the lane from the stile crossed previously a track branches off to the right. This track eventually becomes a footpath which crosses the fields and ends at the sharp bend on the B6049 road about 400 m south of the starting point. From this point there is a fine view of the dolerite quarry at **2**, and on the way back to the car park there are roadside exposures of the Chee Tor Rock to examine. Beware of the traffic on the road; frequently there is a great deal of it.

# 6  Ashford and Monsal Dale

(1 day)

*Purpose*: The exposures seen on this excursion consist of limestones and mudstones near the top of the Carboniferous Limestone (Dinantian) Series. Interbedded with the limestones are two basaltic lava flows.

*OS maps*:     1:50 000 Sheet 119
               1:25 000 The White Peak Outdoor Leisure Map
*IGS maps*:    1:50 000 Sheet 111, Buxton (S or D)
               1:25 000 Sheets SK16, Monyash, and SK17, Miller's Dale

The excursion follows a circular route of about 8 km along public footpaths and minor roads. It can be started equally well from the village centre at Ashford (195698), or at the White Lodge picnic site on the A6 road at 171707, or at Monsal Head (185715). There are parking and toilet facilities at all these points. Refreshments are obtainable at Ashford and Monsal Head. There is a Youth Hostel at Bakewell, 3 km east of Ashford. There are bus services from Bakewell to Ashford and Monsal Head, and from Buxton to Ashford.

## Succession

|  |  | Thickness (m) | Zones |
|---|---|---|---|
| Namurian (mainly mudstones) |  | 120 | $E_1-R_1$ |
| Viséan | Longstone Mudstones | 30 | $P_2$ |
|  | Eyam Limestones | 40 |  |
|  | Monsal Dale Limestones | 200 | $D_2$ |

## Itinerary

If the start is made at Ashford, first cross the River Wye at the ancient and picturesque Sheepwash footbridge, then cross the busy A6 trunk road to the footpath on the far side. Proceed westwards along this path for 300 m or so, noting the outcrops of dark grey, muddy limestones on the left. These are part of the Monsal Dale Limestones and, formerly, were extensively worked in the Ashford area for use as a decorative stone. When

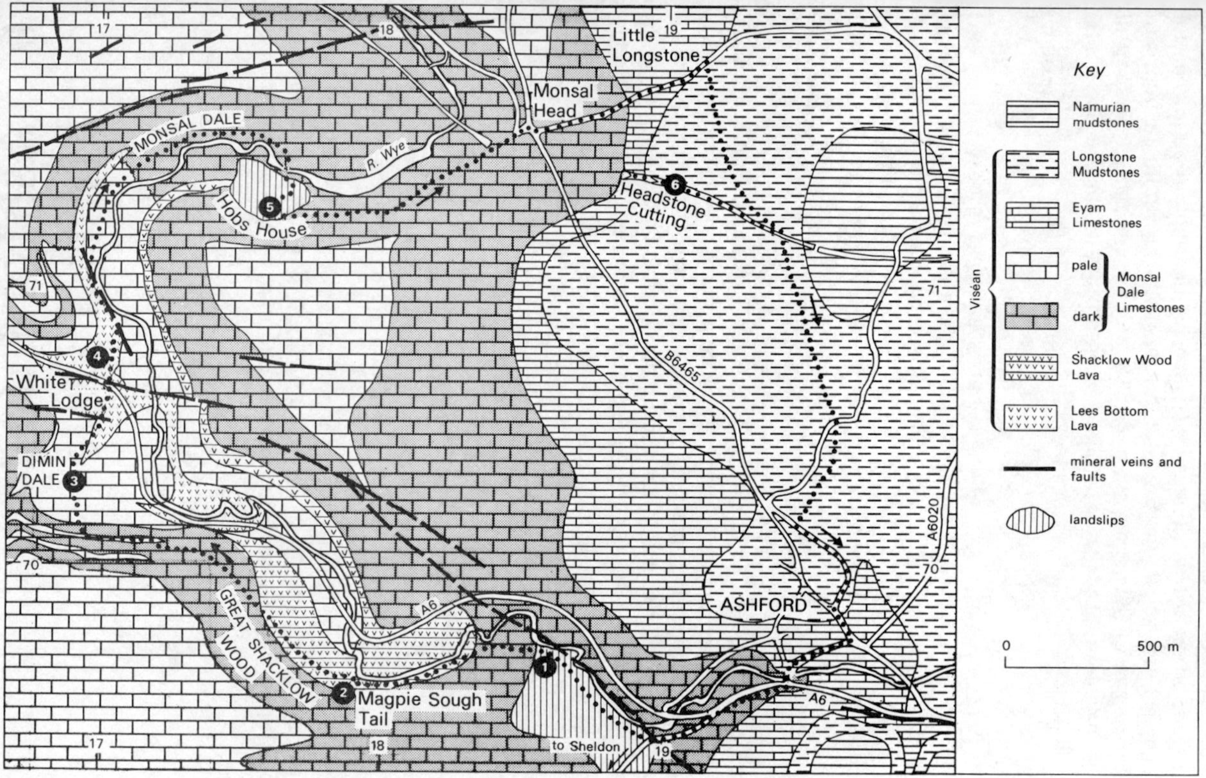
17
18
Little
19
Longstone
Monsal
Head
MONSAL DALE
R. Wye
Hobs House
5
6
Headstone
Cutting
71
4
White
Lodge
DIMIN
DALE
3
B6465
71
70
GREAT SHACKLOW
WOOD
A6
A6020
70
ASHFORD
1
2
Magpie Sough
Tail
A6
17
18
to Sheldon
19
Key
Namurian
mudstones
Viséan
Longstone
Mudstones
Eyam
Limestones
pale
dark
Monsal
Dale
Limestones
Shacklow Wood
Lava
Lees Bottom
Lava
mineral veins and
faults
landslips
0
500 m

**Figure 33** Map of the Ashford and Monsal Dale area.

cut and polished the limestone has a deep black, lustrous finish. It was known in the stone trade as Ashford Black Marble, although in a strictly geological sense it is a limestone and not a true marble.

At 192694 fork left on to a minor road signposted to Sheldon. About 100 m along this road, on the left, there is a disused, and now inaccessible, quarry and mine from which Ashford Black Marble used to be obtained.

At 189694 leave the minor road and follow the riverside path up the main valley. Here, the valley is wide and open with gently sloping sides. Several small outcrops of dark grey, cherty limestones are seen by the riverside at **1** (186697). These are typical of the dark facies of the Monsal Dale Limestones.

Continue up the valley to the Magpie Sough Tail at **2** (179696). At this point water drained from the Magpie Mine, 1.5 km to the south, discharges into the River Wye. The sough was completed in 1881 and, at the tail end, cuts through a basalt flow. The ground in the vicinity of the exit of the sough is littered with loose blocks of the basalt, many of which are vesicular with amygdales of calcite filling most of the vesicles. This basalt flow is the Shacklow Wood Lava, the younger of the two lava flows in the Monsal Dale–Ashford area. Both lava flows are younger than the Miller's Dale Lavas seen in excursion 5.

In Great Shacklow Wood the valley becomes much narrower and deeper. The path through the wood climbs obliquely up the steep, scree-covered slopes before descending to Dimin Dale where it crosses well bedded limestones belonging to the pale facies of the Monsal Dale Limestones. A few silicified brachiopod fossils can be seen in the limestone (**3**) (169703).

At White Lodge cross the A6 again and take the footpath signposted to Monsal Dale. The path soon becomes rather wet and muddy. The reason for this is the presence of the Lees Bottom Lava which, although not well exposed, underlies this part of the area (**4**) (170708). The layer of basalt impedes the natural downward drainage of groundwater with the result that springs emerge along the line of outcrop of the junction between the basalt and the overlying limestones. The spring water, saturated in lime, evaporates to form a precipitate of calcareous tufa, the soft, pale brown, porous rock much in evidence on and around the footpath at **4**.

Continue along the footpath up Monsal Dale. Bedded limestones with much black chert are exposed here and there by the riverside. On both sides of the valley there are steep scree-covered slopes. The scree is a mixture of grey limestone and black chert. Some excellent fossil corals and brachiopods can be found in the scree along with large crystals of calcite.

At 172714 the valley begins to swing round towards the east and the landslip at Hobs House comes into sight on the southern side. Cross the river at the footbridge at 177714 and climb up to the rocks at Hobs House (**5**) (176713) avoiding the area which has been fenced off for conservation

**Figure 34**  Bedded limestone with nodules of black chert at Hob's House.

**Figure 35**  *Dibunophyllum* in cherty limestone at Hob's House (× ½).

**Figure 36** *Lithostrotion* at Hob's House (× ½).

**Figure 37** View of Monsal Dale from Monsal Head.

purposes. The exposures at **5** consist of well bedded, cherty, dark grey Monsal Dale Limestones (Fig. 34). One bed of limestone 2.5 m thick is crowded with fossil corals. The single coral *Dibunophyllum* (Fig. 35) and several species of the compound coral *Lithostrotion* (Fig. 36) are particularly abundant. This bed with its distinctive fossils can be traced round much of Monsal Dale and the neighbouring dales.

Return to the main footpath and proceed to Monsal Head. Here there is one of the Peak District's most spectacular viewpoints (Fig. 37), with the River Wye now deeply entrenched in its narrow, twisting valley.

If time is limited it is possible to return direct from Monsal Head to Ashford by the B6465 road, but this narrow road, with no footpath and heavy traffic, is highly unpleasant for pedestrians. A much more interesting and enjoyable route back follows the minor road from Monsal Head through Little Longstone village. At the eastern end of the village, on the southern side of the road, two footpaths start. Take the one signposted to Ashford. It crosses open meadows and descends to a shallow, gently sloping valley in complete contrast to the deeply incised Monsal Dale so close by.

At 193713 the footpath crosses an abandoned railway line. It is worth making a detour about 400 m along the track to the west to see the Headstone Cutting **(6)** (190713). Here there are excellent exposures of the Longstone Mudstones and the Eyam Limestones. The mudstones, occupying a gently folded syncline, are exposed close to the bridge which crosses the cutting. They are interbedded with thin bands of muddy limestone and contain a rich assortment of fossils including species of goniatites, brachiopods, trilobites and ostracods.

The Eyam Limestones are exposed at the far end of the cutting near the entrance to the Monsal Head tunnel. They consist of fairly massive beds of dark grey, cherty limestone. Some of the beds are finely laminated and show signs of cross-bedding. Minor washouts are evident and in places the lamination is greatly contorted, particularly in areas where there are large concentrations of chert nodules.

From the Headstone Cutting return to the footpath and follow it down through the fields to Ashford.

# 7   Lathkilldale

(1 day or ½ day)

*Purpose*: The gorge cut by the River Lathkill into the rocks of the Carboniferous Limestone (Dinantian) Series is examined. Scenically this is one of the Peak District's most beautiful dales. It also contains much of great interest to geologists.

Bedded limestones of the pale facies of the Monsal Dale Limestones predominate at the eastern end of the dale, but the succession is interrupted twice by flows of basaltic lava. In the central section of the dale the dark facies of the Monsal Dale Limestones is seen and at the western end of the dale there is a quarry in reef limestone at the base of the Eyam Limestones.

For centuries the mining of lead ore was a major industry in Lathkilldale and some interesting relics of this activity remain. Much of the dale is now a National Nature Reserve and is strictly protected. Please observe all the rules relating to access to the reserve. These are posted on the notice boards at the entrances. Hammering of rocks is NOT permitted in the reserve; consequently your hammer should be left at home.

There is a choice of two routes. The longer is about 12 km in length, and in places the going is rough. The shorter route, along easy footpaths and minor roads, is about 8 km long and can be completed in half a day. The dale is at its best in spring and early summer.

*OS maps*:  1:50 000 Sheet 119
1:25 000 The White Peak Outdoor Leisure Map
*IGS maps*:  1:50 000 Sheet 111, Buxton (S or D)
1:25 000 Sheet SK16, Monyash (covers the western half of the area)

The excursion starts from the car park in the village of Over Haddon (204665, point P on Fig. 38). There are toilet and picnicking facilities at the car park. At Bakewell, 3 km to the north, there is a Youth Hostel and other accommodation. There is a frequent bus service between Bakewell and Over Haddon. Refreshments can be obtained in the village.

**Succession**

|  |  | Thickness (*m*) | Zones |
|---|---|---|---|
| Viséan | Eyam Limestone | 40 | $P_2$ |
|  | Monsal Dale Limestones | 100 | $D_2$ |

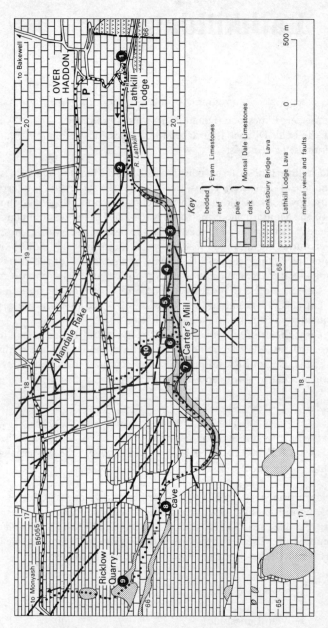

**Figure 38** Map of the Lathkilldale area.

Two extrusions of basalt occur within the Monsal Dale Limestones succession. The lower flow, the Lathkill Lodge Lava, varies from 0 to 4 m in thickness, and the upper layer, the Conksbury Bridge Lava, reaches a maximum thickness of about 30 m.

The strata have not been folded to any significant degree, but they have been dislocated by several minor faults. The fault planes have greatly facilitated the movement of mineralising fluids. Consequently, many of the mineral veins coincide with fault planes.

## Itinerary

From Over Haddon take the lane which winds southwards down to the riverside at Lathkill Lodge (203662). The footbridge which crosses the river at this point is built of stout blocks of limestone containing many large crinoids and brachiopods. During the winter months the river flows strongly under the bridge, but in the summer the river bed is frequently dry at this spot, the river having disappeared underground some distance upstream. On these occasions it reappears 200 m downstream from Lathkill Lodge in the form of copious springs of crystal-clear water.

Before proceeding up Lathkilldale it is worth making a short detour downstream to see the springs and also the Lathkill Lodge Lava at 1 (204662). Small exposures of highly weathered vesicular basalt appear on the side of the dale close to the footpath.

Return to Lathkill Lodge, then follow the footpath along the north bank of the river for another 600 m or so, passing outcrops of grey, cherty limestones bearing scattered corals and brachiopods. Two short trial adits driven into the hillside in search of lead ore are also seen (Fig. 39).

On entering the wooded section of the dale at 197661 the footpath divides. The right-hand path quickly leads to the ruins of the Mandale Mine at 2 (197662). This mine, believed to be one of the oldest in the Peak District, ceased working in 1851, but the remains of the engine house and the pumping shaft are still there. Lead ore was extracted from the Mandale Rake to a depth of 30 m below ground level, but the operations were greatly hindered by the excessive amounts of water encountered underground. A pump installed to remove this water was powered by a large water-wheel. The wheel was driven by water diverted from the river at a point about 1.2 km upstream from the mine and conveyed to it along an aqueduct. The water raised from the mine, along with the water used in the water wheel, was discharged into a sough which drained into the river and the exit of this sough can be seen close to the point where the footpath divides.

The vein of the Mandale Rake is exposed on the side of cliff behind the mine. On no account should the old workings be entered: they are in a very

**Figure 39**   A trial adit in Lathkilldale.

dangerous condition. Specimens of calcite, barite and galena are common on the waste tips between the mine and the river.

About 150 m upstream from the mine the aqueduct crossed the river. The water was conveyed in a wooden trough supported on a series of limestone pillars, some of which still survive in a somewhat dilapidated condition.

At **3** (192658) a large hollow excavated in the ground between the footpath and the river marks the site of another water wheel supplying power to operate pumps for draining the lead mines. This wheel, erected in 1836, had a diameter of 52 feet (nearly 16 m) and was said to be the second largest water wheel in Britain.

Continue along the footpath up the dale for another 300 m or so to **4** (189658). Here, on the hillside above the path, dark Monsal Dale Limestones were once quarried on a small scale for use as a decorative stone. When polished the limestone had a deep black colour and was known in the stone trade as 'Black Marble'. It is not a true marble in the geological sense and is more correctly described as a very fine-grained muddy limestone.

Another 250 m up the dale at **5** (186658) further signs of old mining activities appear. The vein, the Gank Hole Vein, runs obliquely up the side of the dale and was worked in the 1880s for lead ore and ironstone. The iron mineral extracted was limonite, hydrated iron oxide, and it was used to make the yellowish-brown pigment, ochre. Traces of the mineral can be found on the old waste tips by the riverside.

Location **6** (184657), reached just after the footpath emerges from the wooded section of the dale, is the site of Carter's Mill. Not much remains of the mill apart from two hefty millstones lying beside the mill dam. Typically they were made from massive coarse sandstone of the Millstone Grit Series.

The longer excursion route continues up the main valley of Lathkilldale from **6** but the shorter route diverges at this point and ascends the tributary dale on the right to **10** below.

In the main valley at **7** (181657) there is, during most of the year, a

**Figure 40**   Monsal Dale Limestones in Lathkilldale.

waterfall. This is caused by a thick bed of calcareous tufa deposited here by the precipitation of calcium carbonate from river water saturated with lime. The tufa encrusts the leaves and stems of the aquatic plants growing in the river.

Above **7** the dale assumes a much more rugged character. Steep scree-covered slopes with craggy outcrops of limestone line each side of the dale (Fig. 40). The river is now much reduced in size and, in places, the bed is completely dry most of the year. At **8** (171659) there is a large natural cave on the southern side of the dale. The stream which flows out of the cave in wet weather is regarded as the present source of the River Lathkill but, obviously, at one time the source must have been much further to the west to enable the river to excavate the deeply incised valley which extends to Ricklow Quarry and the Monyash area beyond.

At Ricklow Quarry **(9)** (165661) richly crinoidal limestone was extracted for use, when cut and polished, as an ornamental stone. There was a great deal of waste from this operation, so that numerous blocks of stone litter the quarry and its surroundings. The limestone is part of a knoll-reef at the base of the Eyam Limestones. The upper part of the reef is crowded with specimens of the large brachiopod *Gigantoproductus*. These are

**Figure 41**   Cross-sections of *Gigantoproductus* fossils in limestone at Ricklow Quarry (× ⅓).

spectacularly displayed on the bedding planes of the large blocks of limestone displaced from the upper part of the quarry face (Fig. 41).

From **9** follow the footpath northwards through the quarry to join the B5055 road and return to Over Haddon. On the way there are numerous traces of abandoned lead mines to be seen in the fields adjoining the road. Some of the mineral veins are being reworked for fluorite.

The shorter route leaving Lathkilldale at **6** follows a track northwards up an unnamed tributary dale. At several places alongside the track fossiliferous limestones are exposed and at **10** (182660) they are especially rich in brachiopods.

Return to Over Haddon along the minor road. The effects of former lead mining are much in evidence where the road crosses the Mandale Rake.

# 8 Youlgreave and Alport

(1 day or ½ day)

*Purpose*: The exposures seen on this excursion span the upper part of the Carboniferous Limestone (Dinantian) Series and the lower part of the Millstone Grit (Namurian) Series. The succession includes limestones of reef and shelf facies, mudstones, shales and sandstones. Numerous mineral veins traverse the area. These have been extensively worked for lead ore in the past and some are currently being worked for fluorite.

The walking distance is about 8 km along public footpaths and minor roads. If necessary the excursion can conveniently be divided into two half-day excursions of about equal length.

*OS maps*:     1:50 000 Sheet 119
                1:25 000 The White Peak Outdoor Leisure Map
*IGS map*:     1:50 000 Sheet 111, Buxton (S or D)

The excursion starts at All Saints Church in the centre of Youlgreave (212643). There are bus services from Matlock and Bakewell to Youlgreave. Cars may be parked at the side of the playing fields on the Alport road on the east side of the town. There are several hotels, cafés and a Youth Hostel in Youlgreave.

## Succession

|  |  | Thickness (m) | Zones |
|---|---|---|---|
| Namurian | Ashover Grit | over 50 | $R_2$ |
|  | mudstones and shales | 200 | $E_1 - R_1$ |
| Viséan | Longstone Mudstones | 15 | $P_2$ |
|  | Eyam Limestones | 30 |  |
|  | Monsal Dale Limestones | over 75 | $D_2$ |

Subsequent to their deposition the beds were tilted towards the south-east. The angle of dip is generally between 5° and 10°.

## Itinerary

Starting from All Saints Church in the centre of Youlgreave proceed north

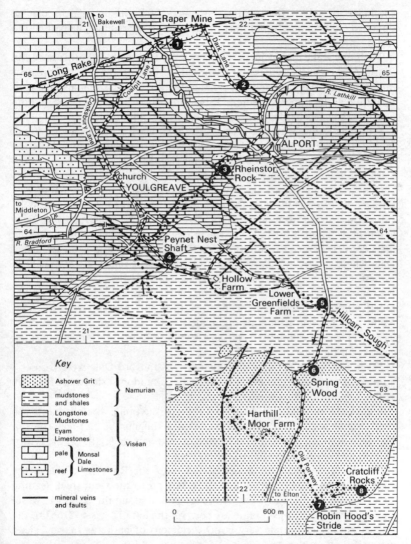

**Figure 42** Map of the Youlgreave and Alport area.

along Conksbury Lane for 250 m or so, and then, just after passing the telephone exchange, turn to right down the narrow Coalpit Lane which leads to a footbridge over the River Lathkill. On the far side of the river take the footpath up a steep slope to **1** (216652) where the Long Rake, one of Derbyshire's most important mineral veins, is exposed. Excellent specimens of yellow fluorite bearing traces of barite and galena can be obtained from an outcrop of part of the vein on the right of the path at the top of the slope.

The Long Rake is coincident with a normal fault with downthrow to the south. At present the vein is being worked for fluorite in a series of opencut mines on each side of Conksbury Lane. Until recently a similar pit was in operation at the Raper Mine near **1**, but work there has now ceased and the quarry is being backfilled with waste from the other mines.

From the Raper Mine site turn south down Dark Lane for about 400 m to **2** (221649). Here black shales are exposed in the bank at the side of the lane. They are part of the Longstone Mudstones at the top of the Carboniferous Limestone Series.

Continue along Dark Lane towards Alport. Just before the junction of the lane with the main road through Alport is reached, large deposits of calcareous tufa are seen on the right. On reaching the main road turn right and proceed along it for 150 m to the bridge over the River Lathkill. On the far side of the bridge turn left on to the riverside footpath up Bradford Dale where, at several places, limestones of the Eyam and Monsal Dale Limestones are exposed. The most spectacular of these is the Rheinstor Rock at **3** (219645). This is part of a patch of massive reef limestone within the Eyam Limestones. On the near vertical cliff by the riverside cross-sections of large brachiopods embedded in the limestone are prominent. Those which were hollow inside are responsible for the numerous small cavities on the face of the cliff.

Several of the mineral veins which cross Bradford Dale were formerly worked for their lead content. Old adits and abandoned shafts are the only visible remains of this activity.

Follow the footpath up Bradford Dale to the point where it meets the road coming south from Youlgreave. This is roughly the half-way stage in the excursion and, if it is being treated as two half-day outings, the return to Youlgreave can be made from here.

For the second half of the excursion cross the River Bradford at the road-bridge and then immediately turn left on to the farm track leading southeastwards to Hollow Farm. About 300 m along the track the waste tips of an abandoned lead mine are seen on the left. This is **4** (216638), the Peynet Nest Shaft. Some very good specimens of barite and clear calcite can be found on these tips.

Continue along the farm track and after passing a fish pond, but before reaching Hollow Farm, turn left up another farm track for 250 m, then turn right and proceed to Lower Greenfields Farm.

During the first half of the eighteenth century the mining of lead ore developed into a major industry in the area around Youlgreave and Alport. Many of the disused shafts to be seen in the fields near Hollow and Lower

Greenfields Farm date from this time. As was usual in the mines of Derbyshire, problems arose because of the excessive amounts of underground water encountered once the mines were driven below the water table. In some cases the water could be drained into the Rivers Bradford and Lathkill, but for the deeper mines this was not possible and so in 1766 a start was made in driving a sough from Hillcarr in Darley Dale. 5 km to the east of Alport, to drain the deep mines into the River Derwent.

The sough, driven deep underground through Namurian shales and mudstones, reached the Greenfields Shaft (**5**) (226635) at a depth of about 60 m below the surface, 21 years later in 1787. From the Greenfields Shaft a network of subsidiary soughs was subsequently driven to link up the various other mines in the area to the main sough and thus lower the water table throughout the district and enable the miners to work deeper sections of the veins. When completed the Hillcarr Sough was the longest in Derbyshire. Every minute it discharged many thousands of gallons of water into the River Derwent and it was navigable underground by small boats all the way from Darley Dale to the Greenfields Shaft. The deep mining of lead ore continued in this area until the early part of the present century. All that remains at **5** is a low mound in a corner of a field where blocks of sandstone cover the top of the shaft.

From **5** join the Alport–Elton road and proceed south to the scarp formed by the outcrop of the Ashover Grit at Spring Wood (**6**) (225631). This grit, consisting of thick beds of coarse-grained reddish-brown sandstone, was, and still is, widely quarried in the area for use as a building stone throughout much of northern England.

Because of the wooded nature of the north-facing escarpment the exposures of the sandstone there are not particularly impressive when compared with those to be seen at **7** and **8**, so it is best to follow the road up to and over the top of the escarpment as far as the entrance to Harthill Moor Farm at 223627, then take the footpath heading half left across the fields to Robin Hood's Stride (**7**) (224622). This footpath (Fig. 43) is part of the very ancient Old Portway, another part of which forms a section of the route described in Excursion 2.

Robin Hood's Stride is a massive tor of gritstone. It dramatically illustrates the effects of prolonged weathering and erosion. Cross-stratification, well displayed in some of the beds, indicates a derivation from a southerly direction (Fig. 44).

From Robin Hood's Stride it is a short walk across to an equally spectacular outcrop of the Ashover Grit at Cratcliff Rocks (**8**) (227623). Here differential weathering and erosion along the bedding and joint planes has carved the outcrop into a series of immense blocks of sandstone. These form a line of high cliffs overlooking the low-lying ground to the south (Fig. 45).

**Figure 43** The Old Portway, a prehistoric track at Robin Hood's Stride.

**Figure 44** Robin Hood's Stride.

**Figure 45**  The Ashover Grit at Cratcliff Rocks.

From **8** return along the Old Portway to Harthill Moor Farm and follow the public footpath which leads through the fields back to Youlgreave.

# 9  Black Rock and Steeple Grange

(½ day)

*Purpose*: On this excursion a traverse along a typical Millstone Grit 'edge' is followed by an examination of richly fossiliferous limestones high in the Carboniferous Limestone (Dinantian) succession.

The total walking distance is about 3 km along public footpaths and trails. There are no problems of access, but along the section of the route which crosses land belonging to the Forestry Commission the collection of specimens is not allowed.

*OS maps*:    1:50 000 Sheet 119
1:25 000 The White Peak Outdoor Leisure Map
*IGS maps*:   1:63 360 Sheet 112, Chesterfield (S & D)
1:25 000 Matlock Special Sheet

The excursion starts at the picnic site at Black Rock (290556). The area is within walking distance of Cromford and Wirksworth. Buses on the route which links Buxton, Bakewell, and Matlock with Derby stop in Steeple Grange village close to Black Rock. At the picnic site there is a large free car park with toilets.

## Succession

|  |  | *Thickness (m)* | *Zones* |
|---|---|---|---|
| Namurian | Ashover Grit | 80 | $R_2$ |
|  | mudstones and shales | 120 | $E_1 - R_1$ |
| Viséan | Cawdor Shale | 10 | $P_2$ |
|  | Cawdor Limestone | 40 |  |
|  | Matlock Limestone | over 30 | $D_2$ |

An anticline which pitches gently towards the south-east is the principal fold structure in the area. The Gang Vein, the most important mineral vein in the area, follows the plane of a normal fault which crosses the area from east to west and has downthrow on its northern side. The distribution of the numerous smaller mineral veins appears to be closely related to the

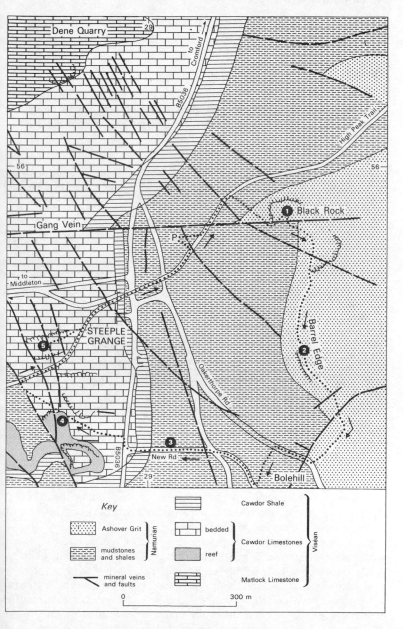

**Figure 46** Map of the Black Rock and Steeple Grange area.

pattern of dominant northeasterly and northwesterly joints in the limestones.

**Itinerary**

From the picnic site (point P on Fig. 46) proceed eastwards along the High Peak Trail for about 100 m and then turn southeastwards along the footpath which leads up to **1** (293557), the conspicuous crag of Black Rock (Fig. 47).

**Figure 47**  The Ashover Grit at Black Rock.

Lead ore was at one time obtained from a mine which worked the Gang Vein near the base of the rock and as a result of this good specimens of galena, sphalerite, barite, fluorite and calcite are plentiful on the waste tip. These specimens belong to the owners of the land, the Forestry Commission, and should not be taken away, but they may be picked up for examination and returned to the tip.

Black Rock, in spite of its name, consists of a light reddish-brown sandstone, the Ashover Grit. At this locality it forms thick beds of cross-bedded, coarse-grained, pebbly sandstone. The cross-bedding appears to be mainly directed towards the north, thus suggesting that the sediments were derived from a southerly source. Because of the very exposed situation of Black Rock the sandstone has undergone severe weathering and erosion (Fig. 48).

From the top of the rock a fine view of the countryside to the north is

**Figure 48**   The effect of weathering on gritstone at Black Rock.

obtained. In Dene Quarry in the foreground well bedded Matlock Limestones dipping gently towards the east are seen. In the distance, to the right, the deep gorge cut in the limestones by the River Derwent at Matlock Bath is a prominent feature.

From Black Rock follow the footpath to the trig point at the summit of Barrel Edge (2) (294553). On the way the path skirts a disused quarry from which sandstone was taken for use as a building stone. Barrel Edge is a perfect example of a Millstone Grit escarpment resulting from the erosion of a gently dipping sequence of soft shales and mudstones overlain by a much harder sandstone. The dip slope is towards the east and the steep scarp faces west. From the view point at the summit the prodigious amount of limestone quarrying that has taken place in the vicinity of Wirksworth is emphatically revealed.

Continue southwards along the footpath following the ridge of Barrel

Edge and descend to meet Oakerthorpe Road on the outskirts of Bolehill village. Turn right and proceed along the road for about 150 m before turning left down a footpath signposted 'The Lanes'. At Bolehill Methodist Church turn right and proceed for 200 m or so to the crossroads just beyond Bolehill Post Office to join New Road (3) (291551). The Ashover Grit features prominently in the construction of many of the houses in the village. In New Road, however, a fine-grained yellowish-grey limestone was used for the kerb. It contains some fine compound corals and many large productid brachiopods.

Proceed down New Road to its junction with the B5036 Wirksworth–Cromford road and cross to the far side. About 30 m up the main road to the right, close to the bus stop, enter a lane which leads to a public footpath passing through the disused Colehill Quarry (4) (288552). Here the Cawdor Limestone is well exposed. The part of the quarry to the north-east of the path is fenced off, but the more interesting south-western part is easily accessible and shows well bedded dark grey limestone overlying and apparently banked up against a mass of pale-coloured reef limestone lacking well defined bedding planes (Fig. 49). The reef limestone contains a rich variety of productid and spiriferid brachiopods, large pinnate bivalves, and the cephalopod *Orthoceras*. In the bedded

**Figure 49**  Bedded limestone overlying reef limestone in Colehill Quarry.

limestone the remains of the brachiopod *Gigantoproductus* and fragments of crinoids are abundant.

From Colehill Quarry continue along the footpath and join the High Peak Trail where it bridges the path. Proceed eastwards along the trail to **5** (288554), Steeplehouse Quarry. The large blocks of limestone lying loose on the floor of the quarry are worth a close inspection. Small cone-shaped fossils, which look rather like tiny limpets, cover some of the bedding planes (Fig. 50). The fossils, some of which are blue-grey in colour, are the dermal denticles (toothlike structures) of a Carboniferous fish, *Petrodus patelliformis*. In places so many of the denticles are present that they give the surface of the bedding planes a rough, dimpled texture.

From **5** return to the High Peak Trail and then to the picnic site at Black Rock.

**Figure 50** *Petrodus patelliformis* in limestone in Steeplehouse Quarry (× 1).

# 10  Thorpe Cloud and Lower Dovedale

(1 day)

*Purpose*: The object of this excursion is to examine the spectacular gorge cut by the River Dove through the basin and reef facies of the Milldale Limestones.

Most of the route is on well maintained public footpaths but some scrambling up and down steep slopes is also involved. The total walking distance is about 8 km.

*OS maps*:        1:50 000 Sheet 119
                      1:25 000 The White Peak Outdoor Leisure Map
*IGS maps*:      1:50 000 Sheet 124, Ashbourne (S or D)
                      1:25 000 Sheet SK 15, Dovedale
(Publication of both the IGS maps is expected in 1982.)

The excursion starts at the car park in Lower Dovedale (146509, point P on Fig. 51), where there are toilets and, during the summer months, refreshment facilities.

## Succession

|  | | *Thickness (m)* | *Zones* |
|---|---|---|---|
| Carboniferous | Widmerpool Formation | 50 | $P_1$–$P_2$ |
| Limestone | Hopedale Limestones | 80 | $D_1$ |
| Series | Milldale Limestones | 300 | $C_1$–$D_1$ |

## Itinerary

From the car park at 146509 proceed northeastwards up the dale road for about 100 m before crossing the river by a footbridge. The huge, pyramidal mass of Thorpe Cloud lies directly ahead. Typical reef limestones are seen almost immediately at the base of the hill (1) (148510) (Fig. 52). The stratification of the rocks is very poorly defined but more or less coincides with the steep sides of the hill. Patches of algal limestone are conspicuous on account of their very fine grain and unusually light grey colour (Fig. 53).

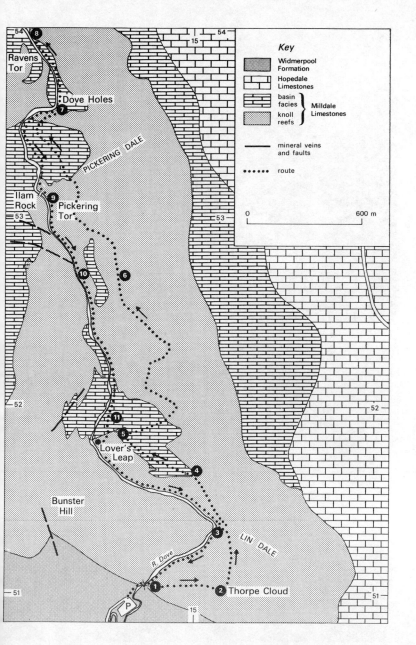

**Figure 51** Map of the Thorpe Cloud and Lower Dovedale area.

**Figure 52** Exposures of reef limestone on Thorpe Cloud.

**Figure 53** Algal Limestone at Thorpe Cloud.

Further exposures of similar algal limestone can be seen at various outcrops as the ridge leading to the summit of Thorpe Cloud is ascended. Location **2** (152510) at the summit provides a magnificent view up Dove Dale and across the valley to Bunster Hill on the far side. Algal limestone is still plentiful at the summit and there are scattered patches in which productid and spiriferid brachiopods are abundant.

From **2** descend along the path leading northwards to the confluence of Lin Dale and Dove Dale. The stepping stones across the river at this point **(3)** (152514) are worth a close look. Polished by many thousands of footsteps, the blocks display interesting cross-sections of brachiopod and crinoid fossils. These show up particularly well when the surface is wet.

**Figure 54**   Cave in the Milldale Limestones at Dove Holes.

From **3** the route follows the footpath which climbs obliquely northwards up the eastern side of the main valley. There are frequent exposures of dark grey limestone, often richly crinoidal, on or close to the footpath near **4** (150517). These belong to the basin facies of the Milldale Limestones and were deposited at about the same time as the algal limestones seen on Thorpe Cloud but, obviously, in a different environment.

Location **5** (147518), a spur of limestone jutting out into the main valley, provides splendid views up and down the dale.

**Figure 55**   Pinnacle of Milldale Limestones at Pickering Tor.

Northwards from **5** the path skirts the edge of the plateau overlooking the main valley and craggy outcrops of reef limestone provide further dramatic views. At **6** (146527), on the eastern side of the path, there is a small disused quarry in limestone rich in productid brachiopods.

The footpath dips down to cross the tributary valley of Pickering Dale, rises again, and then gradually descends to reach the main valley close to the two large natural caves at Dove Holes (**7**) (142536). The caves, which appear to be water-worn, are in massive beds of poorly stratified, pale grey reef limestone (Fig. 54).

Upstream from **7** outcrops of thinly bedded, dark grey, cherty limestones of the basin facies appear on both sides of the valley close to the riverside. Reef limestone appear to overlie the basin limestones here, but on the cliff of Ravens Tor (141539) a lateral transition between reef and basin facies occurs. This is clearly visible from the footpath at **8** (141540).

From **8** return down the dale following the riverside path to **9** (142531). At this point the most conspicuous features are the tall, narrow pinnacles of Pickering Tor and Ilam Rock on opposite sides of the river (Fig. 55). There are several such pinnacles in Dove Dale. All are in reef limestone and why it is that in some parts of the dale the rock should be so resistant to weathering and erosion as to form these near vertical pillars and yet in other parts of the valley it has readily broken down into scree is by no means clear.

Further downstream, at **10** (144527), a spring issues from a small cave beside the footpath. Two closely spaced minor faults can be seen in the roof of this cave and it may be the presence of these faults that allows the groundwater to escape to the surface here.

Further down the dale limestones of the basin facies briefly replace those of the dominant reef facies so that outcrops of bedded, dark grey limestone appear on the footpath at **11** (145519) and again on the steep northern side of the knoll of reef limestone at Lovers Leap (145518).

From Lovers Leap continue down Dove Dale, cross the river either at the stepping stones at **3** or, further downstream, at the footbridge, and return to the car park.

# 11 Hartington, Wolfscote Dale and the Tissington Trail

(1 day or ½ day)

*Purpose*: On this excursion limestones of the shelf, reef and basin facies of the Carboniferous Limestone (Dinantian) Series are examined.

The total walking distance is about 11 km along public footpaths, trails and minor roads. Mostly these are in good condition, but in wet weather some parts of the path in Beresford Dale become very muddy and slippery. If time is limited a shortened version of the excursion involving about 6 km of walking can be undertaken.

*OS maps*:       1:50 000 Sheet 119
                 1:25 000 The White Peak Outdoor Leisure Map
*IGS maps*:      1:50 000 Sheet 111, Buxton (S or D)
                 1:25 000 Sheet SK15, Dovedale

The excursion starts at the village of Hartington (130605), picturesquely situated in the valley of the River Dove about 15 km south-east of Buxton. There are cafés and toilet facilities in the village, and accommodation is available in hotels and a splendid Youth Hostel. Cars may be parked in the centre of the village, but it may prove difficult to find space on busy summer weekends. There are bus services from Buxton (daily except Sundays), Ashbourne (on Thursdays and Saturdays) and Leek (on Saturdays only).

**Succession**

|  |  | Thickness (m) | Zones |
|---|---|---|---|
|  | Monsal Dale Limestones | 150 | $D_2$ |
|  | Bee Low Limestones | 200 | $D_1$ |
| Viséan | Woo Dale Limestones | 100 | $S_2$ |
|  | Iron Tors and Milldale Limestones | 100 | $C_2$–$S_1$ |

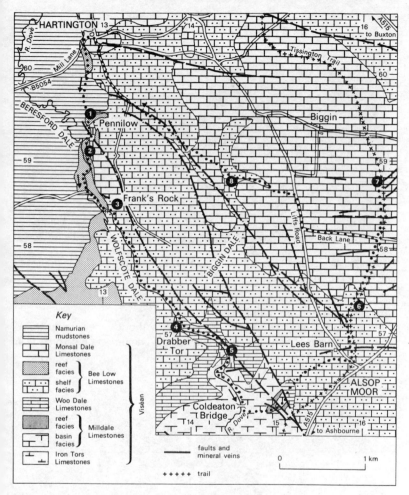

**Figure 56**  Map of the Hartington and Wolfscote Dale area.

## Itinerary

From the centre of Hartington proceed down Mill Lane (the B5054 road to
Warslow) for about 100 m, then turn left on to the footpath signposted to
Beresford Dale. At first this footpath crosses undulating meadowland
dotted with small outcrops of shelly limestone belonging to the marginal
reef facies of the Bee Low Limestones. Location **1** (128595), on the side of
Pennilow, a prominent ridge of reef limestone (Fig. 57), is a typical
outcrop. There is little sign of any bedding in the limestone, but fossils such

**Figure 57** View across Beresford Dale near Hartington. The small hill in the middle distance is Pennilow.

as brachiopods and fragments of crinoids and bryozoa are plentiful. Some goniatites also may be found here.

On the low ground to the west of Pennilow the River Dove meanders over a floodplain of alluvium underlain by soft Namurian mudstones. The

**Figure 58** Frank's Rock at the entrance to Wolfscote Dale.

valley at this point is wide and open. At **2** (128591), where the footpath descends to the riverside, there is a sudden and striking change in the topography as the river enters a deep, narrow gorge cut in reef limestones. For about 500 m downstream there are extensive exposures of these limestones in the near vertical walls of the gorge, then the river, emerging from the gorge, crosses back on to mudstones again and once more the valley opens out wide and flat. This open stretch continues downstream for 300 m or so until at **3** (131584) the river re-enters the limestone area and the valley becomes deep and narrow again (Fig. 58).

Bedded limestones of the shelf facies of the Bee Low Limestones are well exposed at Frank's Rock at **3**. In contrast to the limestones of the same age seen at **1**, the limestones here have clearly defined bedding planes, with a dip northwards of about 10°, and contain relatively few fossils.

The caves at Frank's Rock display fine examples of flowstone formed by the precipitation of calcium carbonate from water percolating through the overlying limestone.

Wolfscote Dale is a narrow and relatively straight V-shaped valley incised to a depth of about 130 m in the limestones (Fig. 59). For most of its length it is cut in Bee Low Limestones but mostly these are concealed beneath extensive scree deposits. The few scattered outcrops show well bedded, light grey limestone with a gentle northerly dip.

**Figure 59**   View down Wolfscote Dale.

At **4**, Drabber Tor (139571), the Woo Dale Limestones appear. These, like the overlying Bee Low Limestones, are thickly bedded, pale grey limestones, but they differ in being slightly harder and, consequently, they produce steep, craggy outcrops of solid rock rather than more gently inclined scree slopes. The crags are conspicuous on both sides of the valley just above the junction of Wolfscote Dale and Biggin Dale.

At this point those taking the full excursion should continue down the main valley, but those on the shorter route should follow the path up Biggin Dale to **8** and return from there to Hartington.

At **5** (144568), 150 m down the main valley from the junction of Wolfscote and Biggin Dales, the Iron Tors Limestones make their appearance. These are thickly bedded, light grey limestones very similar to the overlying Woo Dale Limestones, and, like them, dip northwards at a low angle. Downstream they appear to pass laterally into the darker and much more thinly bedded Milldale Limestones. These also differ in containing many chert nodules.

After crossing the stile just beyond Coldeaton Bridge (146561) turn left up the tributary dale (not named on the OS maps) leading to Lees Barn. On the way outcrops of reefy limestone within the Milldale Limestones form crags close to the footpath.

At Lees Barn bear right along the minor road and join the Tissington Trail at the bridge over this road. Follow the trail northwards for about

**Figure 60**   The Coldeaton Cutting on the Tissington Trail.

500 m. From the high embankment there are splendid views of the surrounding countryside.

At **6** (158571) the trail enters the Coldeaton cutting in which for about 1 km there is continuous exposure of bedded shelf limestones (Fig. 60). The section begins in Bee Low Limestones but the persistent gentle northerly dip eventually brings in the overlying Monsal Dale Limestones. These are more fossiliferous than the Bee Low Limestones and in places are particularly rich in brachiopods and corals. These are specially abundant near the bridge over the cutting. Note, incidentally, that the bridge was constructed with blocks of coarse sandstone and cherty limestone. For some unknown reason the limestone excavated in the cutting was not used in the construction work.

Continue northwards along the trail to the next cutting at **7** (162588). Here the Monsal Dale Limestones are again exposed and, at the northern end of the cutting, they are rich in brachiopod fossils.

From **7** return along the trail, then turn right on to Back Lane and right again on to Liffs Road. About 400 m along Liffs Road a footpath on the left leads down to **8** (145587) in the upper part of Biggin Dale. This very attractive dale is a superb example of a dry valley. As a consequence of a lowering of the water table the stream which formerly flowed down the dale dried up and all the drainage is now underground.

From **8** follow the signposted track straight back to Hartington.

# 12    Ecton and the Manifold Valley

(1 day)

*Purpose*: Rocks belonging to the basin and reef facies of the Carboniferous Limestone (Dinantian) Series are seen on this excursion. In addition the relics of former mining of copper ore on a large scale are seen at Ecton.

The total walking distance is about 6 km along public footpaths, tracks and minor roads.

*OS maps*:    1:50 000 Sheet 119
             1:25 000 The White Peak Outdoor Leisure Map
*IGS map*:    1:50 000 Sheet 111, Buxton (S or D)

If you have your own transport the most suitable place to start this excursion is at Wettonmill (095561), where the National Trust provides ample free parking space, toilets and a café. There are bus services from Buxton (daily except Sundays) and Leek (on Wednesdays and Saturdays) calling at the village of Warslow (087587), and from there it is a walk of about 1 km to Ecton (096585) where the route can be joined.

**Succession**

|  |  | Thickness (*m*) | Zones |
|---|---|---|---|
| Namurian | mudstones and thin sandstones | over 100 | $E_1$ |
| Viséan | Mixon Limestone and Shales | 150 | $P_1-P_2$ |
|  | Ecton Limestones | 170 | $D_1$ |
|  | Milldale Limestones | over 100 | $C_2-S_2$ |

The major structural feature of the area is an anticline with a north–south axis. Dips up to 80° are recorded on its eastern limb. Additionally there are numerous complex minor folds to be seen. These are particularly well developed in the Ecton Limestones.

**Itinerary**

From the riverside car park at 094561 cross the bridge to Wettonmill. Location **1** (095562) is the conspicuous crag of Nan Tor rising behind and slightly to the north-west of the mill. It consists of light grey, cavernous reef limestone within the Milldale Limestone sequence (Fig. 62). The view

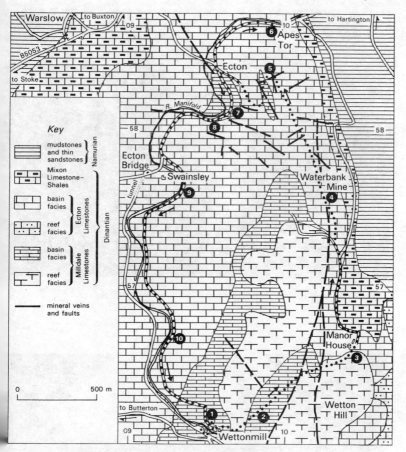

**Figure 61** Map of the Ecton and Manifold Valley area.

down the Manifold Valley from the caves on Nan Tor makes the climb up to the crag worth while (Fig. 63).

From Wettonmill the path runs eastwards obliquely up a hillside for about 150 m before descending to join a path up the tributary dale (not named on the OS maps) coming into the main valley from the north-east.

At **2** (098562) the first of a series of outcrops of dark grey, thinly bedded, crinoidal limestones is exposed beside the footpath. This is the basin facies of the Milldale Limestones and in addition to crinoids the beds contain numerous terebratulid, productid and spiriferid brachiopods. Similar outcrops appear at intervals up the dale for the next 400 m or so.

Wetton Hill, on the south side of the dale, consists largely of knoll-reef

**Figure 62** Cavernous reef limestone at Nan Tor, Wettonmill.

**Figure 63** View of the Manifold Valley from Nan Tor.

limestone. Fore-reef limestones full of crinoids and small brachiopods were deposited on the flanks of the knoll-reef and can be seen in the banks of the small stream at **3** (104565).

Continue up the dale to Manor House, then follow the lane from there northwards to 104572 where the lane forks. Take the left fork up the hillside for about 70 m before turning right on to a public footpath (at present not signposted). Follow this path through fields for about 250 m to the site of the disused Waterbank Mine at **4** (103576).

The Waterbank Mine produced copper, lead and zinc ores at intervals during the eighteenth and nineteenth centuries from veins extending from the surface to a depth of about 50 m. Traces of chalchopyrite, malachite, galena and sphalerite can still be found on the waste tips. Here, and elsewhere on Ecton Hill, the mine shafts are frequently not fenced off, or at best are protected by only a flimsy fence. It is advisable, therefore, to keep well clear of these shafts.

From **4** continue northwards along the footpath, passing several long disused mines and shafts before arriving at **5** (099584). The building here, now a barn, was erected in 1788 to house a Watt steam engine which provided power to raise ore and pump water from the Ecton Mine. The shaft from the engine house was over 400 m deep and, at the time of its construction, it was the deepest mine in Britain. From **5** there are several paths leading to various abandoned mine workings. It depends on the time available how many of these can be examined but a careful search of the waste tips will yield good specimens of the green copper minerals, malachite and chrysocolla, along with much calcite.

Descend to the road along the valley bottom and proceed along it to the disused roadside quarries at Apes Tor (**6**) (099586). In this famous section bedded limestones of the basin facies of the Ecton Limestones, thrown into a series of tightly folded anticlines and synclines, are exposed (Fig. 64).

Return along the road down the valley to **7** (096581). This is a large roadside quarry excavated in a scree of limestone which, in places, has become cemented to form a breccia. The limestones which provided the material for the scree are the Ecton Limestones. They have also been quarried at **8** (095581), about 150 m further along the road. Here thinly bedded, dark grey limestones interbedded with thin partings of shale are seen to be tightly folded into anticlines and synclines. Nodules of black chert are conspicuous in the limestone beds. From **8** a good view of the Ecton copper mines is obtained (Fig. 65).

Continue down the valley road towards Ecton Bridge but just before reaching the bridge branch left on to the minor road which keeps to the eastern bank of the river. It is much more pleasant walking on this road than on the busier road on the west side. Further outcrops of the Ecton Limestones are seen at various roadside exposures all the way along

**Figure 64** Anticlinal fold in Ecton Limestones at Apes Tor.

**Figure 65** The abandoned copper mines at Ecton.

between Ecton Bridge and Wettonmill. The strong folding which the beds have undergone is evident from the constantly changing direction and amount of dip displayed by the strata. Fossils are not common, but brachiopods and crinoids can be found in some of the beds.

The entrance to another old copper mine, the Swainsley Mine, is seen by the roadside at **9** (094576). Very little copper ore was found here and the mine was very wet. The water, however, is now put to good use as part of the public water supply. *Do not on any account enter the mine; it is very dangerous.*

The final exposure **(10)** (093567) again displays contorted beds of cherty limestones similar to those seen at **6** and **8**.

# 13 Hen Cloud, The Roaches and The Ramshaw Rocks

(1 day)

*Purpose*: The rocks seen on this excursion are sandstones and mudstones of Upper Carboniferous age. They belong to the upper part of the Millstone Grit (Namurian) and the lower part of the Coal Measures (Westphalian) Series. They have been folded along a north–south synclinal axis. There is a close correspondence between the topography of the area and its geological structure. The gritstone ridges provide some superb panoramic viewpoints and so, if possible, a clear, sunny day should be chosen for this trip.

The walking distance on this excursion is about 13 km and the route follows public footpaths and minor roads.

| | |
|---|---|
| *OS maps*: | 1:50 000 Sheets 118 and 119 |
| | 1:25 000 The White Peak Outdoor Leisure Map |
| *IGS maps*: | 1:50 000 Sheet 111, Buxton (S or D) |
| | 1:25 000 Sheet SK06, The Roaches & Upper Dove Valley |

As the route is circular, the excursion may be started at the village of Upper Hulme (013610), or at the footpath leading up to Hen Cloud (006615), or at Roach End (996645). There is limited parking space at all of these points. Buses on the Sheffield–Newcastle-under-Lyme route via Stoke, Leek and Buxton (service 208) stop at Upper Hulme. Refreshments are obtainable in Upper Hulme. There is a Youth Hostel at Meerbrook (989608).

## Succession

| | | Thickness (*m*) | Zones |
|---|---|---|---|
| Westphalian | mudstone and thin coal seams | 100 | |
| | Woodhead Hill Rock (sandstone) | 30 | A |
| | mudstone | 30 | |

| Namurian | | | |
|---|---|---:|---|
| | Rough Rock (sandstone) | 40 | G |
| | mudstone | 50 | |
| | Chatsworth Grit | 75 | |
| | mudstone | 40 | |
| | Roaches Grit | 100 | $R_2$ |
| | mudstone | 60 | |
| | Five Clouds Sandstone | 100 | |
| | mudstones and sandstones | 500 | $E_2-R_2$ |

Strong folding along a north—south synclinal axis (the Goyt Syncline, also seen on excursion 15) has caused the strata to dip at angles of up to 40° along each limb of the fold. The syncline plunges to the north at about 10° and, consequently, the outcrops diverge northwards and close towards the south.

### Itinerary

From Upper Hulme proceed westwards along the minor road which climbs steeply through the village. The steep slope, which is due to the relatively resistant Five Clouds Sandstone upon which the village is built, eases off when the overlying mudstone is reached. At 006615 turn right on to the signposted footpath for Hen Cloud and ascend the massive escarpment formed by the Roaches Grit.

On the summit of Hen Cloud (1) (008617) beds of coarse-grained feldspathic sandstone are exposed. They contain a scattering of quartz pebbles up to 2 cm in diameter. The strata dip north-eastwards at about 20° and are strongly cross-bedded. The overall colour of the rock is reddish-brown. On close inspection it is seen to consist mainly of angular quartz grains interspersed with grains of feldspar partly decomposed to powdery clay with a pale pink colour.

From 1 head northwards across a valley to the Roaches and take the very good footpath along the crest of the ridge. The Roaches Grit is exposed nearly all the way along and, as on Hen Cloud, dips strongly towards the north-east (Fig. 67). At Doxey Pool (2) (004629) a small patch of muddy peat overlies the sandstone. This impedes the normal free drainage of rainwater through the sandstone and allows the small pond to form.

Craggy outcrops of the sandstone at 3 (002636) display some of the best examples of cross-bedding to be seen in the area. They indicate derivation from a southerly direction (Fig. 68).

From the trig point at the summit of the ridge (4) (002639) the syncline is seen to be opening out to the north, and, as the strike of the sandstone swings round to a more northwesterly trend, the dip appears to slacken off somewhat and the topographic feature becomes less prominent.

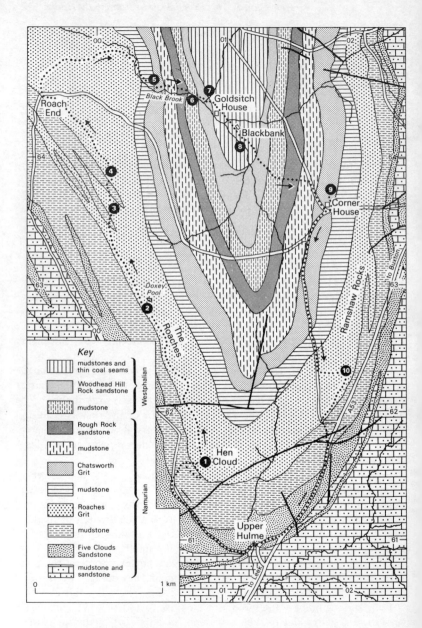

**Figure 66** Map of the Hen Cloud, Roaches and Ramshaw Rocks area.

Key

Westphalian
- mudstones and thin coal seams
- Woodhead Hill Rock sandstone
- mudstone

Narmurian
- Rough Rock sandstone
- mudstone
- Chatsworth Grit
- mudstone
- Roaches Grit
- mudstone
- Five Clouds Sandstone
- mudstone and sandstone

0    1 km

**Figure 67** The Roaches viewed from the top of Hen Cloud.

**Figure 68** Cross-bedding in the Roaches Grit.

Continue northwards along the ridge and, after crossing the minor road at Roach End, proceed down a farm track into the valley of the Black Brook passing several outcrops of the Roaches Grit on the way. Outcrops of black mudstone are seen on the far bank of the stream just before the footbridge over it is reached. The coarse, pebbly sandstones which overlie the mudstone can most easily be examined at **5** (004645) by the side of the footpath 200 m further up the valley. They dip steeply towards the east and are similar to the Roaches Grit in appearance. They belong to the Chatsworth Grit.

Continue along the footpath up the valley. The Rough Rock sandstone is not well exposed here and has only a minor effect on the topography. The mudstones overlying the Rough Rock are seen, however, in the bank of the stream at **6** (007645) just beside the footpath.

From **6** the footpath is not signposted and not well defined, but, heading in the general direction of Goldsitch House, it leaves the streamside and crosses some fields. At **7** (009645) weathered black shale is exposed in the bank of a small tributary stream 30 m upstream of where it is bridged by the footpath.

Follow the path past Goldsitch House and Blackbank farmhouse. Although clearly indicated on the OS maps as a public right of way, the footpath is obviously not used very much. From the footpath further exposures of black shales can be seen in the banks of the Black Brook at **8** (012641). Here, close to the axis of the syncline, the beds dip northwards at 10°. This, in effect, is the plunge of the syncline.

Continue along the path, such as it is, between the fields (at 014639 an ineffective barbed wire fence has to be negotiated) until it meets the minor road at 017638, then turn right along this road. The ridge of high ground on the left is formed by the Chatsworth Grit. Outcrops of the grit can be examined at **9** (018637) beside the track which runs northwards along the

**Figure 69**    View of The Ramshaw Rocks.

ridge from Corner House. The beds dip towards the west at 40° and consist of coarse-grained, cross-bedded sandstone.

Return to the road and proceed south to the Ramshaw Rocks at **10** (019623). Of all the outcrops of the Roaches Grit this is the most spectacular (Fig. 69). Superbly sculptured by weathering and erosion, the massive beds of coarse sandstone, tilted at a steep angle, dominate the skyline with dramatic effect. From the summit of the ridge the view across the axis of the syncline to the corresponding outcrops on Hen Cloud and The Roaches is equally exciting.

From Ramshaw Rocks return along the A53 road to Upper Hulme.

# 14 Parsley Hay, the High Peak and Tissington Trails

(1 day)

*Purpose*: The route of this excursion is along disused railway tracks which have been converted to trails for use by walkers and cyclists. Several deep cuttings through shelf limestones provide excellent exposures of the Carboniferous Limestone (Dinantian) Series, and from the high embankments which alternate with the cuttings there are splendid views of the surrounding countryside.

The total walking distance is about 11 km. There are no problems about access to the trails, but there are restrictions on the use of hammers.

*OS maps*:  1:50 000 Sheet 119
          1:25 000 The White Peak  Outdoor Leisure Map
*IGS maps*:  1:50 000 Sheet 111, Buxton (S or D)
          1:25 000 Sheet SK16, Monyash

The excursion can be started at any of the former railway stations at Hurdlow (127660), Parsley Hay (147637) and Hartington (149612). There is free parking space at all three points. There are toilet facilities at Parsley Hay and Hartington. Refreshments can be obtained at Hurdlow. There is a Youth Hostel in the village of Hartington.

Pedal cycles, available for hire at Parsley Hay on certain days between late March and early November, provide a novel and pleasant alternative to walking. Details of the times and prices of this service can be obtained from the Peak Park Joint Planning Board's office at Aldern House, Baslow Road, Bakewell.

On Saturdays throughout the year a bus service links Hartington with Buxton and Leek. During the summer months there is also a weekend service from Derby and Ashbourne.

**Succession**

Limestones of the shelf facies of the Carboniferous Limestone Series

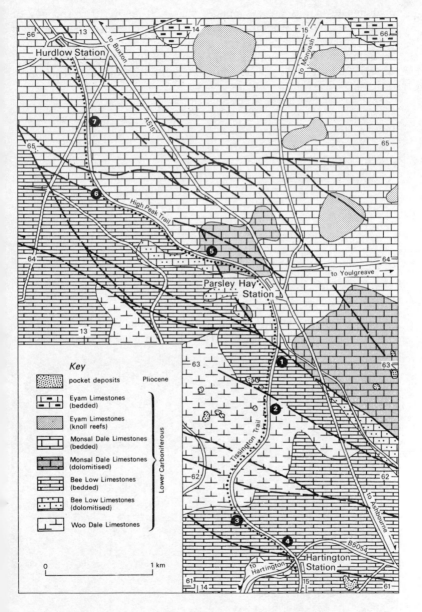

**Figure 70** Map of the Parsley Hay area.

predominate in this area. In places they have been either partially or completely altered to dolomite. The limestones have been gently folded into an anticline with its axis running in a north-north-easterly direction. Several normal faults with a generally NW–SE trend cross the area. Most of these faults are mineralised and some have been worked for lead ore in the past.

Solution hollows in the limestones contain pockets of loose yellow sand and clay. These have been used to make refractory bricks. They are of Pliocene age.

Details of the succession are as follows:

|  |  | *Thickness* (*m*) | *Zones* |
|---|---|---|---|
| Pliocene | pocket deposits | | |
| Lower Carboniferous (Viséan) | Eyam Limestones | 30 | $P_2$ |
| | Monsal Dale Limestones | 100 | $D_2$ |
| | Bee Low Limestones | 120 | $D_1$ |
| | Woo Dale Limestones | 60 | $S_2$ |

**Itinerary**

From Parsley Hay station proceed southwards along the High Peak Trail. After 400 m or so the trail divides. Take the right-hand fork, signposted to

**Figure 71**  Cutting in the Bee Low Limestones south of Parsley Hay.

Ashbourne and Hartington. After another 200 m the trail enters a deep cutting in Bee Low Limestones (**1**) (147631). At first the beds dip gently northwards but the dip soon eases off and along most of the section the beds are practically horizontal (Fig. 71). Fossils are not plentiful here but a few crinoid ossicles and compound corals occur at the northern end of the cutting.

An unexposed fault with downthrow to the north-east crosses the trail near the southern end of the cutting, so that the next exposures, seen at **2** (146625), belong to the Woo Dale Limestones. In contrast with the previous locality fossils are abundant here. Cross-sections of thick-shelled brachiopods of the genus *Gigantoproductus* are conspicuous, along with colonial corals and crinoids.

Silica sand was formerly obtained from several small pits in the area west of the trail at **2**. The sand, mixed with clay, occupied solution hollows in the limestones and is of Pliocene age. Regrettably the pits are now used as rubbish tips.

From **2** the trail runs along an embankment before entering another cutting at **3** (144618). The first exposures consist of hard, pale grey Woo Dale Limestones but the gentle southerly dip soon brings in the overlying Bee Low Limestones. A well defined bedding plane marks the junction between the two formations (Fig. 72).

**Figure 72** Bee Low Limestones overlying the harder and more massive Woo Dale Limestones at **3**. The junction is indicated by a broken line.

Near the southern end of this cutting the limestone has undergone partial dolomitisation with the result that patches of light brown crystalline dolomite occur here and there in the unaltered grey limestone.

In the final cutting on this section of the trail the Bee Low Limestones are again exposed. Grey limestone with brachiopods and crinoid ossicles dips southwards at a low angle (4) (147614). Many of the joint faces in the limestones are coated with crystals of the dog-tooth variety of calcite.

The picnic site at Hartington station is a convenient spot to halt for a rest before returning to Parsley Hay and tackling the northern section of the trail.

Location 5 at 141640 is a short cutting in the Bee Low Limestones. In places they show patchy dolomitisation so that the well bedded pale grey limestone containing shell and crinoid fragments is replaced in places by brown crystalline dolomite with numerous cavities, some of which are filled with secondary calcite (Fig. 73). All traces of bedding planes and fossils disappear in the dolomitised patches.

**Figure 73**   Partially dolomitised Bee Low Limestones at Parsley Hay.

In the next cutting (6) (132646) a fault separating the Monsal Dale and Bee Low Limestones has facilitated the passage of mineralising fluids with the subsequent deposition of thick veins of coarsely crystalline calcite and a little barite.

The Monsal Dale Limestones are seen again at 7 (131652). They resemble the Bee Low Limestones in their light grey colour and profusion of shell and crinoid fossils but differ in containing numerous irregular nodules of chert.

Lead ore was at one time mined in this area and traces of the disturbance caused by the mining can be seen in the fields on each side of the trail between **7** and Hurdlow station (Fig. 74).

From Hurdlow station return to Parsley Hay station.

**Figure 74**  Ground disturbed by lead miners at Hurdlow.

# 15 Goyt's Moss

(½ day)

*Purpose*: The Upper Carboniferous rocks of this area are examined. The succession includes sandstones, mudstones, shales and coal seams. They have been gently folded along a north–south synclinal axis – the Goyt Syncline.

Goyt's Moss is an area of open moorland dissected by a network of small streams forming the headwaters of the River Goyt. It lies about 5 km west of Buxton, just north of the Buxton–Macclesfield road (A537). The walking distance on the excursion is about 5 km over public footpaths and tracks. There are no problems of access, except that on certain weekdays during late August and September parts of the moors may be closed to the general public to allow the shooting of grouse to take place. Details of the closures are posted at the National Park Information Centres and at the access points to the moors.

| *OS maps*: | 1:50 000 Sheet 119 |
| | 1:25 000 The White Peak Outdoor Leisure Map |
| *IGS maps*: | 1:50 000 Sheet 111, Buxton (S or D) |
| | 1:25 000 Sheet SK07, Buxton |

The excursion starts at the Goyt's Moss picnic site at 018716 (point P on Fig. 75), where there is a large free car park with toilets. On Saturdays there is a bus service to the area (alight at the Cat and Fiddle Inn) from Buxton and Macclesfield.

## Succession

| | | Thickness (m) | Zones |
|---|---|---|---|
| Westphalian | sandstone | 5–15 | |
| | grey shale | 5–10 | |
| | Yard Coal | 1 | |
| | Woodhead Hill Rock | 15–30 | A |
| | dark grey shale | 20–40 | |
| Namurian | Rough Rock | 25–50 | $G_1$ |
| | shale | 50–70 | |
| | Chatsworth Grit | over 30 | $R_2$ |

## Itinerary

In the roadside cuttings in the vicinity of the picnic site and in the banks

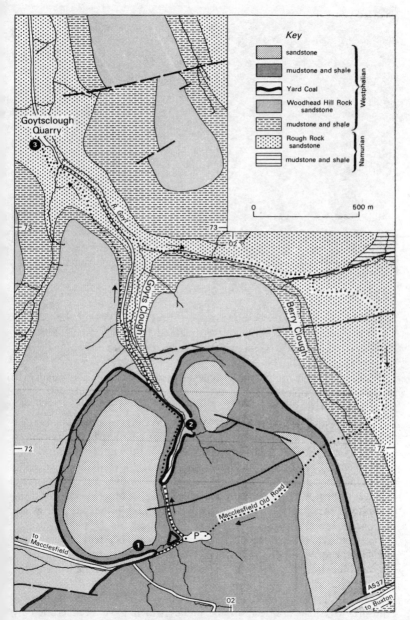

**Figure 75** Map of the Goyt's Moss area.

**Figure 76** Exposures of the Coal Measures at Goyt's Moss.

the nearby stream there are abundant exposures of the Westphalian shales.
The best section is in the banks of the stream at **1** (016715), where dark grey
shale overlies the Yard Coal which in turn rests on a bed of pale grey
ganister (Fig. 76).

A careful search in the dark shales 1–2 m above the coal seam will
provide fine examples of the fossil bivalve *Carbonicola* and, with luck,
some scales of fossil fishes.

The coal seam, which is just above the level of the stream, was once
mined here, and what is now seen are pillars of coal left by the miners to
support the roof of the mine while they extracted the intervening portions
of the seam. It is a bright, hard coal, but of fairly low quality.

The ganister beneath the coal was originally a sandy subsoil underneath
the layer of peat which eventually became the coal seam. The roots of some
of the trees penetrated into the subsoil and are now preserved in it as the
fossil *Stigmaria* (Fig. 6).

The ganister grades downwards into a well bedded, fine-grained
sandstone, the Woodhead Hill Rock. This is exposed in the stream bed at
several points upstream from **1**. The sandstone here is dipping towards the
east at about 10°.

From **1** return to the road and proceed down Goyt's Clough. Numerous
exposures of shale are seen in the gullies which have been eroded by
tributary streams draining into the main valley. The tail end of a sough

which drains some of the old mines in the Yard Coal can be seen on the far bank of the stream at a point where there is a footbridge across it (2) (018721). The Woodhead Hill Rock is also exposed here. It is so strongly cross-bedded that it is difficult to determine the precise direction of dip of the strata.

Continue down the road for about 400 m, then follow the footpath on the left signposted to Goytsclough Quarry. At the quarry, (3) (011735) the Rough Rock sandstone was once worked for use as a building stone. It is a coarse-grained feldspathic sandstone containing scattered pebbles of quartz. At the southern end of the quarry the sandstone is cross-bedded. The direction of this bedding is generally towards the south-west, thus implying a derivation from a northeasterly point.

A feature of the Rough Rock throughout the South Pennines is the occurrence within it of large concretions. They are nearly spherical in shape

**Figure 77**   The Rough Rock in Goytsclough Quarry showing a spherical concretion.

and they range in size from a few centimetres to over a metre in diameter. They are composed of the same sandy sediment as the rest of the Rough Rock and, significantly, the bedding planes in the ordinary sandstone continue unbroken through the concretions. This indicates that the concretions are structures which have developed in the rock subsequent to its deposition.

Typical concretions are exposed in the main part of the quarry at Goyt's Clough (Fig. 77). The rusty brown colour of the weathered surfaces of the concretions may be due to the oxidation of iron-bearing minerals present in the cementing agent in the concretions.

From **3** rejoin the road and return back up the main valley for about 400 m before branching off to the left on the footpath signposted to Berry Clough and Buxton. Berry Clough is a tributary valley along the outcrop of the soft shales in between the Rough Rock and the Woodhead Hill Rock. There are few exposures of solid rock in this part of the area but the views across the moorland are impressive. At 026727 the path divides; take the right fork. This leads up to the crest of a ridge formed by the Rough Rock. At 027723 the rock is exposed in a shallow quarry where stone was taken for wall and road construction. The sandstones dip to the west at 10°.

Turn right on reaching the Macclesfield Old Road and head back towards the picnic site. The road, built in 1759 as the Buxton–Macclesfield Turnpike, was replaced by the present main road in 1821. Strongly cross-

**Figure 78** Goyt's Moss, once a busy coal-mining area.

bedded sandstones, exposed on the surface of the road near the junction with the footpath from Berry Clough, are part of the Rough Rock. There is a dip in the road as it crosses Berry Clough, where grey shales are seen in the banks of the stream. The road then rises again over the ridge formed by the Woodhead Hill Rock and the unnamed sandstone above it. On the dip-slope on the far side of this ridge several shallow mines once worked the Yard Coal, but the only visible remains of these mines are some water-filled hollows where the shafts were and some grass-covered mounds marking the sites of the old waste tips (Fig. 78).

# Appendix I   Glossary of geological terms

**algae**   A group of primitive plants; some of the marine forms developed a limy skeleton which is often preserved in fossil remains.

**amygdales**   Rounded aggregates of secondary minerals deposited inside vesicles in rocks of volcanic origin.

**anticline**   A fold in rocks such that the fold surfaces are arched upwards.

**barite**   A vein mineral composed of barium sulphate.

**basalt**   A fine-grained rock of volcanic origin and composed mainly of feldspar and iron silicates.

**basin limestone**   Dark grey, well stratified, impure limestone thought to have been deposited in somewhat deeper water than **shelf limestone**.

**bivalves**   Animals, such as mussels, belonging to a class of the phylum Mollusca, in which the animal is protected by two shells or valves.

**boulder clay**   A deposit formed beneath an ice sheet or glacier and consisting of a mixture of clay, sand, pebbles and boulders.

**brachiopods**   A phylum of marine animals having an external skeleton consisting of two shells or valves; the valves are usually unequal in size.

**bryozoans**   A phylum of small aquatic animals most of which are colonial in habit and share a common supporting skeleton composed of calcium carbonate.

**bullions**   Hard concretions found in shales and mudstones; they consist of clay cemented with calcium and iron carbonates.

**calcareous**   Composed of, or containing, calcium carbonate.

**calcite**   The commonly occurring, crystalline form of calcium carbonate.

**chalcopyrite**   Copper iron sulphide, the principal ore of copper.

**chert**   A very fine-grained form of silica, similar to flint.

**chrysocolla**   A hydrated silicate of copper with a strong blue-green colour.

**clay wayboards**   Layers of decomposed volcanic ash which occur interbedded with limestones in the Peak District.

**cleavage plane**   Flat, and usually lustrous, surfaces developed on many, but not all, crystalline minerals when they are broken; the property of cleavage is largely determined by the way in which the atoms and molecules of the mineral are arranged.

**corals**   Marine animals belonging to the phylum Coelenterata; the majority secrete a hard skeleton composed of calcium carbonate and this is frequently preserved as a fossil.

**crinoids**   Marine animals belonging to the phylum Echinodermata; they have a complicated skeleton composed of calcium carbonate. In most forms the skeleton consists of a body, or calyx, arms and a stem with which the animal was attached to the sea floor. Fossilised fragments of crinoid skeletons occur very commonly in Carboniferous limestones.

**cross-stratification** Inclined layers within a stratified bed of, usually, sandstone, and formed by the deposition of sand grains on a sloping surface such as the front of large underwater ripples or dunes.

**dolerite** A medium-grained igneous rock composed largely of feldspars and iron silicates; it is generally found in small intrusions.

**dolomite** A mineral composed of calcium magnesium carbonate occurring sometimes as a primary constituent of limestones but often as a secondary replacement of **calcite**.

**facies** All the features displayed by a sedimentary rock which enable its environment of deposition to be deduced.

**fault** A fracture in rocks along which movement has taken place displacing the rocks on one side of the fault relative to those on the other side.

**flowstone** A deposit of calcium carbonate coating the sides of caves and fissures in limestone strata.

**fluorite** A vein mineral composed of calcium fluoride; it is also known as fluorspar.

**flute marks** Moulds of marks made by the scouring action of water flowing over newly deposited mud and silt, and preserved as a negative imprint on the base of the next layer of sediment to be deposited.

**galena** A vein mineral composed of lead sulphide; the principal ore of lead.

**gastropods** Animals forming a class of the phylum, Mollusca; most have a univalve shell coiled in a helicoid spiral. They include the snails and limpets.

**goniatites** Marine animals belonging to the ammonoid group of the cephalopod class of the Mollusca; they have a univalve shell coiled in a plane spiral.

**groove marks** Narrow elongated ridges found on the underside of beds of sedimentary rocks; formed in much the same way as **flute marks** by the movement of sediment-laden water over the surface of the underlying layer of sediment before it was consolidated into a rock.

**head** A deposit of clay mixed with sand and angular boulders; it was formed by the weathering of exposed rocks in areas adjacent to ice sheets and glaciers and subject to prolonged and repeated freezing and thawing.

**hydrothermal mineralisation** The deposition, in fissures and cavities in rocks, of minerals which had been transported there in the form of hot, aqueous solutions.

**karstic weathering** The type of weathering that causes the formation of cave systems and **sinkholes** in limestone areas; it is a consequence of the solubility of limestone in slightly acid water.

**lithology** The distinctive features of a rock, such as its texture, structure and composition.

**load casts** Marks on the underside of a bed of sedimentary rock caused by its being forced down into the surface of the underlying bed at a time before either of the beds had become hard, consolidated rocks.

**malachite**   A carbonate and hydroxide of copper with a distinctive bright green colour.

**mudstone**   A rock formed from muddy sediment by compression and consolidation.

**olivine-basalt**   A type of **basalt** in which olivine, an iron magnesium silicate, is one of the principal constituent minerals.

**oolitic**   A texture displayed by some limestones and ironstones and said to resemble fish roe in appearance; oolitic rock is composed of ooliths, minute spherical grains closely packed together to give the rock a fine, uniform texture.

**ostracods**   Small aquatic animals belonging to the class, Crustacea; each ostracod was enclosed by two **calcareous** shells which are frequently found as fossils.

**productid**   Belonging to a group of **brachiopods** characterised by the genus, *Productus*; they are now extinct, but were very common in Carboniferous times.

**rake**   A large mineral vein filling a near vertical fissure in limestone strata; rakes can be up to 7 m wide, several kilometres long and have been mined to depths of over 200 m without reaching bottom.

**reef limestone**   Poorly stratified or unstratified limestone composed largely of the fossilised remains of reef-building plants and animals such as **algae** and **corals**.

**rhynchonellid**   Belonging to a group of **brachiopods** characterised by the genus, *Rhynchonella*; they were common in Carboniferous times and a few species survive to the present day.

**shale**   A fine-grained sedimentary rock which can be split into thin, wafer-like sheets parallel to the bedding surfaces.

**shelf limestone**   Pale grey, stratified limestone thought to have been deposited in shallow, clear water covering a relatively stable part of the sea floor.

**sill**   An intrusion of igneous rock formed by magma being forced between layers of pre-existing sedimentary rocks and so producing a sheet-like structure parallel to the original bedding.

**sinkhole**   A steep-sided depression formed on the surface of areas underlain by limestone as a consequence of solution of the rocks taking place underground.

**sole marks**   Marks such as **flute marks**, **groove marks** and **load casts**, found on the underside of beds of sedimentary rocks.

**sphalerite**   A vein mineral composed of zinc sulphide.

**spiriferid**   Belonging to a group of **brachiopods** characterised by the genus, *Spirifer*; the group is now extinct but numerous species existed in Carboniferous times.

**stalactite**   Pendant structures which hang down from the roofs of caves and are formed by the slow precipitation of calcium carbonate from dripping water.

**stalagmite**   Columnar structures on the floors of caves; they are formed by the precipitation of calcium carbonate from water dripping down from the roof of the cave.

**syncline**   A fold in rocks in which the fold surfaces are trough-shaped.

**terebratulid**   Belonging to a group of **brachiopods** characterised by the genus,

*Terebratula*; the group was common in Carboniferous times and still survives at the present time.

**throw** The amount of displacement in a vertical sense as a result of movement of a **fault** in rocks.

**toadstone** A Derbyshire name for **basaltic** volcanic rocks which have undergone decomposition to some degree.

**turbidite** Sediment, usually of a sandy nature, laid down rapidly in a part of a sea or lake where currents are causing conditions to be turbulent.

**unconformity** A surface, within a sequence of sedimentary rocks, formed by erosion during a break in the deposition of the sequence.

# Appendix II Further reading

Readers interested in the full details of the geology of the northern part of the Peak District are referred to the Memoir of the Institute of Geological Sciences entitled *The geology of the country around Chapel en le Frith (explanation of one-inch Geological Sheet 99)* by I. P. Stevenson and G. D. Gaunt, published by HMSO.

Papers dealing with specific aspects of the geology of the Peak District appear regularly in scientific journals such as the *Mercian Geologist* and the *Proceedings of the Yorkshire Geological Society*.

The Peak District Mines Historical Society have published several fascinating books and papers relating to early mining activities in the area and, jointly with the Peak Park Planning Board, they have published *Lead mining in the Peak District*, an excellent book which is highly recommended to all with an interest in Peak District geology.

Most of the fossils whose names are mentioned in this guide are illustrated in *British Palaeozoic Fossils* published by the British Museum (Natural History).

# Index